Managing Crises in Tourism

Acolla Lewis-Cameron · Leslie-Ann Jordan ·
Sherma Roberts
Editors

Managing Crises in Tourism

Resilience Strategies from the Caribbean

palgrave
macmillan

Editors
Acolla Lewis-Cameron
University of the West Indies
St. Augustine
Trinidad and Tobago

Leslie-Ann Jordan
University of the West Indies
St. Augustine
Trinidad and Tobago

Sherma Roberts
University of the West Indies
Cave Hill, Barbados

ISBN 978-3-030-80237-0 ISBN 978-3-030-80238-7 (eBook)
https://doi.org/10.1007/978-3-030-80238-7

This Palgrave Macmillan imprint is published by the registered company Springer Nature Switzerland AG
The registered company address is: Gewerbestrasse 11, 6330 Cham, Switzerland

Foreword

Islands as Elsewhere

In the western imagination, warm water islands are always 'elsewhere' (Bonnett, 2020), a 'pleasure periphery', exotically situated at the fringe (Ash & Turner, 1976). They cannot be too close: they would otherwise risk becoming infected by the dross of the routine of the cold and grey of daily life that characterizes much of the winters of Western Europe and North America; yet they also cannot be located too far so as to be unreachable except in one's wildest dreams. For the millions of inhabitants of such islanded places as the Caribbean, however, these same islands are home. For most of their inhabitants, there may very well be nowhere else. And so, what may be a disappointing vacation cancellation to a wannabe tourist, may well mean the loss of work or employment to a Caribbean islander engaged in the hospitality industry.

The Specific Challenges of Small Island Jurisdictions

As I write (mid-April, 2021), the Soufriere Volcano has been spewing ash over the north end of the island of St Vincent, the main and most populated island of the Caribbean country and small island developing state (SIDS) known as St Vincent and the Grenadines (SVG). News agencies have reported that some 16,000 people have been obliged to leave their homes. One begs the question, however, as to where such people are being evacuated. The island of St Vincent is 29 km long and 18 km wide: a strong enough volcanic eruption can quickly send its whole population—about 110,000—into emergency mode, obliging a comprehensive evacuation off island. The destination of any evacuated Vincentians is not obvious: their country is independent and no other sovereign state is obliged to take on SVG's populace. Its neighbouring SIDS—St. Lucia, Grenada, Barbados, Antigua & Barbuda and Jamaica have however offered support. This is unlike a similarly named volcano on a similarly small Caribbean island—Montserrat, a UK overseas territory and not a state—which erupted in 1995. Then, that other Soufrière's eruptions had rendered more than half of that island territory uninhabitable, destroying the capital city, Plymouth, and obliging extensive evacuations: some two thirds of the population had left the island, many relocating not to neighbouring Caribbean island states, but to the motherland and colonial power: the UK.

Such situations help to frame the specific challenges that confront small island states and territories when faced by massive environmental or economic shocks. Were a similar eruption to unfold on a larger and continental country, individuals can be relocated and rehoused somewhere else on a sprawling land area. It is safe to assume that any 'disaster zone' can be cordoned off, while its residents are shepherded to safety in another part of the same jurisdiction. Not necessarily so in a small jurisdiction: when such disasters strike, they tend to affect the whole resident population and its whole land area. The few who can and have the means to do so—aptly referred to by Sheller (2013) as the 'kinetic elite'—would sail or fly away (if endowed with advance warning), possibly on their own boats or private jets. The rest of the resident population has to lump it, dig in, stay put and hope for the best, including benefitting from expressions of international solidarity. Any solace and assistance are more likely to be requested by those subnational island jurisdictions (SNIJs) that remain parts of larger and richer states. Here, patron states—Colombia, France, Mexico, the Netherlands, the UK, the USA and Venezuela, in the case of the island Caribbean—would come demonstrably to the help of their co-nationals, issuing largesse in what are often highly publicized actions. Sovereignty in such situations, in contrast, means that the affected small island state must approach and lobby the international community to provide support, and with no guarantees that such would materialize, and if so, whether on time and in the form that is most valued (Schultz et al., 2019).

I welcome this book to the ever-expanding stock of literature written by the Caribbean, about the Caribbean, and using frameworks and templates that make sense to the Caribbean as a small island region. Here is another valid contribution to a select body of knowledge that boldly seeks to better understand the predicament of Caribbean small states and territories. It does so effectively by deploying a methodology that critically examines 'the dilemma of tourism overdependence' in what continues to be the best known and most strongly branded tourism playground region of the world.

Overdependence: Recipe for Disaster

That overdependence is a recipe for disaster is self-evident. When a major industry suffers, then the whole economy can go into nosedive: this is a manifestation of vulnerability. When the suffering eases, the economy 'bounces back': and this becomes an expression of resilience. Such dynamics are aggrandized and enhanced, the smaller the size and population of a jurisdiction. No surprise, therefore, that small island states insist that their vulnerability to external shocks is an ingrained and structural weakness.

Yet, even in the 99^{+}% of the time that the small island economy is not under the effects of a devastating volcanic eruption, earthquake, hurricane, flood, tsunami or any other natural or human-made mishap, the powerful economic lobby that drives its major industry is often locked in tight embrace with the political elite and ensures that its interests are protected since it is 'too big to fail'. Let us not forget that the absence of economies of scale and of a large and competitive domestic market may mean that various services, some essential, that are delivered in small island states or territories are done so invariably by monopolistic service providers: energy, communication, sea and air transport, banking ... there may be a long list of companies that command no domestic rivals in small economies. Thus, powerful forces are at work to convince one and all that a dominant economic situation is not to be challenged. As a result, when the negative effects of such a dominance are exposed in the wake of an economic or environmental shock, the political narrative is carefully directed and massaged towards 'bouncing back' rather than thinking hard about alternatives and trying to prevent a similar future shock from having similarly devastating consequences.

This is why many countries do not learn from past mistakes. Even when major industries are thumped and pounded by events and trends, the kneejerk reaction is to re-establish the *status quo ante*: restoring major industries to their previous perch and status as, well, major industries. This is obviously a prequel for yet further shocks. And the cycle of vulnerability and resilience continues. And yet, this scenario is often presented as a paradigm: by definition, it allows no other interpretation of the state of affairs. As a consequence, moments of crisis are not also looked upon and considered as serendipitous episodes of opportunity to reconstitute, or at least reconsider, the economic fabric of a small country.

Which is why *building back better* is smarter advice and a more useful interpretation of resilience than merely bouncing back (Rhiney, 2020; Collodi et al., 2021). And we interpret *building* here beyond the literal meaning of the word: as in, building 'hurricane proof' houses (Hallegatte et al., 2018; Barclay et al., 2019). A measure of this smart reconfiguration of the economic set-up of a small island state and territory must also engage with the relative erosion of significance of those sectors and industries on which these economies are *too* reliant.

Tourism: Posterchild of Overdependence; Victim of its Own Success

The poster child of over-reliance in the island Caribbean is tourism. The industry has grown and, in the space of a few decades, has basically transformed Caribbean island economies into tourism platforms and their ancillary services. It has grown in significance also because other traditional economic sectors—bananas, sugar, some manufacturing—have collapsed or lost market share (Fridell, 2011; Seleka & Dlamini, 2020). Tourism remains an industry that has melded so well with the representations of culture and nature of the Caribbean that one has become synonymous with the other. It remains an industry which has emerged and matured without the need for small island state stakeholders and governments in the region to argue for its protection, for subsidies or for favourable market access. It has gripped various island economies so tightly that the locals feel obliged to protest not just about constrained access to their own beaches; but even about being cheated of *views* of those beaches (e.g. Mycoo, 2006).

With the onset of COVID-19, and the subsequent restrictions imposed on international travel, the Caribbean region is now smarting from its own tourism successes throughout the years. Hotels, restaurants, taxi and self-drive services, farmers and fishers, self-catering accommodation providers, tour operators … all these have seen a massive decline in tourism demand for their products or services since Spring 2020. Some relief may have come about through the resort by locals to 'staycations' and/or fiscal stimulus and cheques issued by the state to companies under duress or laid off workers; but any per capita revenue here pales in comparison with international tourism receipts. Thus, the island states and territories whose economies have been most dependent on tourism are amongst the ones that have suffered the most in the post-COVID-19 epoch (Murphy et al., 2020). But the corollary also holds true: for countries like Trinidad and Tobago, and especially Trinidad—since Tobago is heavily dependent on tourism—where tourism is important but where the economy is more diversified and differentiated—thanks mainly to a strong manufacturing base and its suppliers—then, COVID-19 economic suffering has been less severe. Which would explain why we have a Tobago-based chapter in this collection; but none on Trinidad.

Smart, Strategic, Flexible Diversification

I would like to dwell here on the mitigating measures that many citizens of small island states and territories naturally gravitate towards when faced with calamities of the kind being faced today in the island Caribbean. After all, it is not a question of if but when: in the last two decades, the region has had to contend with quite a plethora of crises: the economic consequences of the 9/11 terrorist attacks in the USA; the 2007–2008 financial crisis; various devastating hurricanes; earthquakes; the coronavirus pandemic; (and volcanic eruptions). More, no doubt, will be forthcoming. In such a context, individuals, families and households are naturally savvy and resourceful in a streetwise kind of way: none wishes to get caught napping and rendered destitute and adrift, without any fallback positions for eking a livelihood. And so, emerges an interesting but sustainable 'parallel economy' as 'crisis management': a tourism sector may loom large at a macro, national level; but there is a common, canny resort to various strategies of diversification at a micro (individual, family or household) level. It is as if people opt to take those smart and sound measures that make so much sense, providing 'insurance' against mishap; even as their governments, possibly locked into inaction by powerful lobbies, take no such initiatives to diversify. Here is a crisis-proofed version of resilience, a portfolio deployed and resorted to by island communities 'as they contend with the dynamism of their external and internal environments' (Baldacchino, 2011).

These smart and sound measures include: (1) combining a private sector with a public sector job or contract: the latter is not dependent on tourism; (2) maintaining links with members of one's diaspora overseas, since such members of one's extended family could be counted upon to provide remittances or loans in times of need, and especially when major investment projects are contemplated; (3) spending some time oneself abroad, working elsewhere and saving enough money to stash for that rainy day, which will come; (4) petty self-employment, including part-time farming, fishing, herding, animal husbandry and odd craft, trade and construction work: these add variety to the spice of island life and ensure multiple (albeit limited) revenue streams; and (5) informal and grey economy activity, such as higgling, often undertaken under the radar of formal employment, and thus not captured by tax agencies.

Options (2) and (3) above are more significant in the cases of small island territories. The link with the metropole is a prized resource, and probably the most important benefit enjoyed by citizens of small island territories (unlike those of SIDS), and the main factor in thwarting any aspirations for their independence. 'Securing double residencies, undergoing multiple stints overseas for work and education, tolerating complex transportation routes and itineraries, becomes the order of the day, every day, for those who have the resources and capabilities of doing so; and the dreams and desires of those who cannot' (Baldacchino 2020, 123).

Thus, the same hotel housekeeper, restaurant waitress and taxi driver who are currently out of work because tourists have disappeared in the age of COVID-19, somehow ensure for themselves a decent income from non-tourism operations. Such operations are probably easier to set up and nurture for the non- and peri-urban segment of the population. This blend of formal and non-formal occupations is predicated on and is itself a nudge towards, a resort for strategic, flexible specialization (Baldacchino, 2019). This trait has long been acknowledged in the Caribbean (e.g. Comitas, 1964).

Two Diversification Strategies

Diversification strategies follow two complementary paths. The first involves diversification of the tourist product *within* tourism. Steering away from the bland 'sun, sand and sea' package and venturing instead towards the expansion and consolidation of particular niches—such as agro-tourism, carnival events or scuba diving—help the tourism industry to branch out 'away from the beach': engage with different market segments, tap into different local resources, and connect with different aspects of the local community (e.g. Cuffy, 2017). Cruise tourism has come on stream as such a new market niche, bringing senior citizens from Canada and the USA to the Caribbean; but their actual added value to local island economies is disappointingly limited (Chen et al., 2019).

The second path is to diversify *beyond* tourism. To do so effectively requires a combination of forward thinking, shrewd investments and, admittedly, some strokes of good luck. A number of countries have developed financial service sectors whose low tax advantages appeal to the rich citizens of other countries: but such practices have increasingly fallen within the purview of larger states who are loathed to see their tax revenue base contract (Vlcek, 2017). Controversial 'citizenship by investment' schemes were pioneered in the Caribbean: they offer naturalization and 'passports for sale' to those wealthy and keen enough on a back-up plan; but their association with shady characters tends to lead to serious reputational damage for the countries that dabble with them (Shachar, 2018). A few states and territories have natural and mineral resources that they can export bauxite from Jamaica; avocados from St Lucia; oil and gas from Trinidad; mace and nutmeg from Grenada; and aloe from Aruba. But selling such products without further processing at home does not accrue significant profit and does not create much stable employment. In contrast, the legacy of cultural icon and reggae artist Bob Marley on Jamaica is profound, enduring and integral to that island brand (Johnson & Gentles-Peart, 2019). The creative industries in particular hold considerable potential in the Caribbean region (Nurse, 2009).

The coronavirus pandemic unleashed in 2020 has provided an opportunity for a timely and sober reassessment of tourism, at much the same time that concerned stakeholders—including island citizens—were daring to question the wisdom of the mass tourism algorithm and were also emboldened enough to speak plainly about 'overtourism', possibly for the first time (Milano et al., 2019). The excessive strain on the natural environment, the depletion of non-renewable resources, the monopoly power of service providers … even the very dominance of the industry over the small Caribbean island economy, has cast tourism under a spotlight of intense scrutiny. Once the symbolic reset button is pressed, and life returns to some sense of 'normal' post-COVID-19 lockdowns and quarantines, will life, and island tourism, be the same as they were pre-pandemic? And, if not, what changes can we expect, and on whose initiatives will they unfold? These are the pressing questions of our times.

Conclusion

To conclude, let me return to St Vincent and the Grenadines. The volcanic eruption of April 2021 grabbed the news headlines for a few days (until the death was announced of Prince Philip, Duke of Edinburgh). Of course, for SVG, there is a beckoning future. Past the short burst of international visibility, SVG may yet have nudged an interest and curiosity in the broad community for it to emerge as a stronger tourism destination post-eruption (and post-COVID-19). At least one other island state has managed to grow its tourism economy *fourfold* after a volcanic eruption that disrupted air travel in Northern Europe back in 2010: Iceland.

I invite the readers of this book to delve into its chapters and share in its sober assessments of how one can take some ownership over one's vulnerability and critically envision a resilient and sustainable tourism in, and for, the 'elsewheres' of the island Caribbean.

Godfrey Baldacchino
President, International Small Islands Studies Association (ISISA), Editor, Small States & Territories Journal
Professor of Sociology, University of Malta, Msida, Malta

References

Ash, J., & Turner, L. 1976. *The golden hordes: International tourism and the pleasure periphery*. New York: St. Martin's Press.

Baldacchino, G. 2020. How it feels 'on the ground': The experiences of residents from a subnational island jurisdiction. *Small States & Territories* 3(1): 123–126.

Baldacchino, G. 2019. Competitiveness of small states: insights on flexible specialisation. *Small States & Territories* 2(1): 41–54.

Baldacchino, G. 2011. Surfers of the ocean waves: change management, intersectoral migration and the economic development of small island states. *Asia Pacific Viewpoint* 52(3): 236–246.

Barclay, J., Wilkinson, E., White, C. S., Shelton, C., Forster, J., Few, R., ... & Honychurch, L. 2019. Historical trajectories of disaster risk in Dominica. *International Journal of Disaster Risk Science* 10(2): 149–165.

Bonnett, A. 2020. *Elsewhere: A journey into our age of islands.* Chicago IL: University of Chicago Press.

Chen, J. M., Petrick, J. F., Papathanassis, A., & Li, X. 2019. A meta-analysis of the direct economic impacts of cruise tourism on port communities. *Tourism Management Perspectives* 31: 209–218.

Collodi, J., Pelling, M., Fraser, A., Borie, M., & Di Vicenz, S. 2021. How do you build back better so no one is left behind? Lessons from Sint Maarten, Dutch Caribbean, following Hurricane Irma. *Disasters* 45(1): 202–223.

Comitas, L. 1964. Occupational multiplicity in rural Jamaica. In *Proceedings of the AmericanEthnological Society*, 1963, pp. 41-50. Retrieved from http://cifas.us/pdf/Comitas%20Pdf%20Files/Short%20Manuscripts/1964a%20Occupational%20Multiplicity%20in%20Rural%20Jamaica.pdf.

Cuffy, V. 2017. Carnival tourism. In S. Agarwal, G. Busby & R. Huang (Eds.), *Special interest tourism: Concepts, contexts and cases*, 97–111. Wallingford UK: CAB International.

Fridell, G. 2011. The case against cheap bananas: lessons from the EU-Caribbean Banana Agreement. *Critical Sociology* 37(3): 285–307.

Hallegatte, S., Rentschler, J., & Walsh, B. 2018. *Building back better: Achieving resilience through stronger, faster, and more inclusive post-disaster reconstruction.* Washington DC: World Bank.

Johnson, H., & Gentles-Peart, K. (Eds.). 2019. *Brand Jamaica: Reimagining a national image and identity.* Lincoln NE: University of Nebraska Press.

Milano, C., Cheer, J. M., & Novelli, M. (Eds.). 2019. *Overtourism: Excesses, discontents and measures in travel and tourism.* Wallingford UK: CAB International.

Murphy, M. M., Jeyaseelan, S. M., Howitt, C., Greaves, N., Harewood, H., Quimby, K. R., ... & Hambleton, I. R. 2020. COVID-19 containment in the Caribbean: the experience of small island developing states. *Research in Globalization* 2: 100019.

Mycoo, M. 2006. Sustainable tourism using regulations, market mechanisms and green certification: a case study of Barbados. *Journal of Sustainable Tourism* 14(5): 489–511.

Nurse, K. 2009. *The creative sector in CARICOM: The economic and trade policy dimensions.* Kingston, Jamaica: CARICOM Secretariat.

Rhiney, K. 2020. Dispossession, disaster capitalism and the post-hurricane context in the Caribbean. *Political Geography* 78: 102171.

Seleka, T. B., & Dlamini, T. S. 2020. Competitiveness of ACP sugar exporters in the global market. *The International Trade Journal* 34(2): 247–277.
Shachar, A. 2018. The marketization of citizenship in an age of restrictionism. *Ethics & International Affairs* 32(1): 3–13.
Sheller, M. 2013. The islanding effect: post-disaster mobility systems and humanitarian logistics in Haiti. *Cultural Geographies* 20(2): 185–204.
Shultz, J. M., Kossin, J. P., Shepherd, J. M., Ransdell, J. M., Walshe, R., Kelman, I., & Galea, S. 2019. Risks, health consequences, and response challenges for small-island-based populations: Observations from the 2017 Atlantic hurricane season. *Disaster Medicine and Public Health Preparedness* 13(S1): 5–17. Retrieved from https://discovery.ucl.ac.uk/id/eprint/10045771/1/Shulz_Risks_Health_Consequences.pdf
Vlcek, W. 2017. *Offshore finance and global governance*. New York: Palgrave Macmillan.

Acknowledgements

We wish to acknowledge the contributions of our fellow authors in this volume, for their patience and cooperation in bringing this publication to fruition. We hope that this was an insightful and educational exercise that will hopefully spur additional research on tourism resilience and sustainable tourism development in the Caribbean. This is needed now more than ever! We hope that they find our editorial edits acceptable and all errors and mistakes remain our responsibility.

We are sincerely grateful to the staff at Palgrave Macmillan for taking on this project and for their kind assistance during the preparation and submission of this manuscript.

We are also grateful to our families for their continued love, support, encouragement and assistance for putting up with us during this period. They are to be commended for giving us the necessary time and space to work, given the added pressures of online teaching and school, quarantine and lockdowns and the general uncertainty of the time.

It is our hope that this volume will ignite future research in this area and provoke innovative solutions, as we contend for a more sustainable Caribbean future.

Contents

Part I Background and Conceptual Context

1 **Introduction** 3
Acolla Lewis-Cameron

2 **Understanding Small Island States and Territories** 13
Acolla Lewis-Cameron

3 **Conceptualizing Resilience in Small Island States** 33
Sherma Roberts

Part II Environmental Resilience

4 **Implications for Resilience in the Cruise Tourism-Marine Protected Area Nexus in the Caribbean: The Case of St. Lucia** 67
Myrna Ellis

5 **Tourism Resilience in the Caribbean Island of Cozumel: Best Practice and High-Risk Areas** 89
Kennedy Obombo Magio

6 **A Resilient Eco-Tourism Island: A Case Study of Dominica and Its Tourism Recovery Strategies Post 2017 Hurricane Maria** 109
Tenisha Brown-Williams and Amanda Charles

Part III Socio-Cultural Resilience

7 **Community Resilience in the Face of Natural Disaster: Puerto Rico's Adventure Tourism Industry** 135
Mechelle N. Best and José H. González

8 **An Integrated Path Towards a Resilient Tourism Sector in North-East Tobago** 155
Joanna Moses-Wothke, Aljoscha Wothke, and Leslie-Ann Jordan

9 **Grenada: A Vision of Integrated Technological Advancements to Build a Resilient Tourism Future Through Youth Involvement and Consumer-Centric Service Excellence** 175
Kimberly Thomas-Francois, Ye Shen, and Marion Joppe

Part IV Economic Resilience

10 **Increasing the Resilience of Micro, Small and Medium Tourism Enterprises to Tropical Cyclones in Small Island Developing States** 197
Thalia Balkaran and David Smith

11 **Building Resilience by Strengthening the Link Between Tourism and Agriculture: An Assessment of the Purchasing Patterns of Selected Hotels and Guesthouses in Jamaica** 221
Tolulope Bewaji, Eritha Huntley Lewis, Clive Scott, and Twila-Mae Logan

12 **Collaborative Tourism Entrepreneurship as a Community Resilience Strategy: A Case Study of Castara, Tobago** 241
Shinelle Smith and Leslie-Ann Jordan

Part V Contemporary Landscape & Reflections

13 **COVID-19 Policy Responses—The Tension Between Lives and Livelihoods in Tourism-Dependent Caribbean Territories** 271
Acolla Lewis-Cameron and Sherma Roberts

14 Conclusion 289
Sherma Roberts, Acolla Lewis-Cameron, and Leslie-Ann Jordan

Index 299

Notes on Contributors

Thalia Balkaran is a Trinidadian who recently completed her Ph.D. in Environmental Management at the University of the West Indies, Mona Campus. She has conducted research and has been involved in varying projects in many of the small islands of the Caribbean region focused on the topics of vulnerability, sustainable livelihoodsand disaster risk reduction. Her hope is to apply the knowledge she has gained thus far to helping island states around the world increase their resilienceto disaster events.

Mechelle N. Best Barbadian by birth, Caribbean and global in outlook, Dr. Mechelle Best is a Professor in the Department of Recreation and Tourism Management at California State University Northridge. She is a former Fulbright/Organization of American States Scholar for the Eastern Caribbean whose research interests include cultural heritage, parks and protected areas, and sustainability in tourism. Her current research streams focus on the intersection between Caribbean slavery heritage and tourism and Caribbean tourism's resilienceto natural disasters. Asa life-long lover of the outdoors, Mechelle is also interested in how of people of African descent in the USA experience the outdoors.

Tolulope Bewaji graduated from Temple University with a Ph.D. in General and Strategic Management. She also obtained an M.B.A. from University of Louisiana at Monroe. She taught at St. Xavier University and the University of the West Indies. She is currently a Lecturer at the University of the West Indies. She has published papers related to entrepreneurshipand minority topics.

Tenisha Brown-Williams holds a B.Sc. in Tourism and Hospitality Management (Honours), an M.Sc. Tourism Development and Management (with Distinction) from the University of the West Indies (UWI) and is currently completing her Ph.D. in Governance at the Sir Arthur Lewis Institute for Social and Economic Studies, UWI (St. Augustine). Tenisha has been lecturing to hospitality and tourism students for over a decade and has also worked in the Caribbean tourismsector in product development, sustainable tourismand tourism quality assurance. Her research interests extend to tourism governance, the tourism sharing economy, tourism education and tourism recovery and resilience.

Amanda Charles heads the Sustainable Tourism Division at the Caribbean Tourism Organization (CTO). In this capacity, she leads the organization's efforts in designing and implementing projects, also supporting advocacy and capacity buildingaimed at enhancing the growth, competitiveness and sustainability of Caribbean tourism. A Tourism professional with over 15 years' experience working in tourism development, Ms Charles, has worked at several regional and international organizations, including the Association of Caribbean States (ACS) and Organization of American States (OAS). Ms. Charles holds a Bachelor's Degree in Tourism and Business Management and a Master of Tourism Administration in Sustainable Destination Management. She is from the Caribbean island of Dominica.

Myrna Ellis is a Ph.D. Candidate at the University of the West Indies, Jamaica. Her research area is Sustainable Development with a tourism focus. She also holds Master's degrees in Environmental Science & Management and Tourism Development & Management. Presently, employed as a Senior Lecturer at the Caribbean Maritime University in Jamaica, she coordinates the Bachelor's degree in Cruise Shipping and Marine Tourism. Building on her natural science foundation, her research interests include: cruise tourism, marine protected areas, sustainable development, blue economy, carrying capacity, crisis managementand mixed methods research. Ms. Ellis has also co-authored two publications.

Dr. José H. González is an Assistant Professor in the Department of Recreation and Tourism Management at California State University

Northridge. Born and raised in Puerto Rico, he has 25 years of professional experience in outdoor recreation and adventure programming. He has co-authored two books and several articles for the canopy tour and Zipline industry. He is an active instructor trainer for the American Canoe Association with several certifications in sea kayaking and stands up paddleboarding. As a sailor, he has sailed in the USA, USVIs, BVIs, Leeward Islands and the Windward Islands. He maintains his USCG Captain's License.

Marion Joppe, Ph.D. is a Professor in the School of Hospitality, Food and Tourism Management, University of Guelph, Canada. She specializes in sustainability with a focus on policy, destination planning, development and marketing, and the experiences upon which destinations build. She has extensive private and public sector experience, having worked for financial institutions, tour operators, consulting groups and government, prior to joining academia.

Leslie-Ann Jordan is a Senior Lecturer, Hospitality and Tourism Management at the University of the West Indies (UWI), Department of Management Studies, St. Augustine Campus, Trinidad. She holds a B.Sc. in Tourism Management (First Class Honours) from the University of the West Indies and a Post-Graduate Diploma (with Distinction) and Ph.D. in Tourism Policy and Planning from the University of Otago, New Zealand. Leslie-Ann is also a Certified Manager of Quality/Organizational Excellence (CMQ/OE) from the American Society for Quality (ASQ) and holds a Professional Certificate in Event Management, George Washington University and Institute of Business (IOB). She is the lead editor of the text, '*Sports Event Management: The Caribbean Experience*' which was published by Ashgate, London. Her research interests include tourism development in small island developing states (SIDS) with special reference to the Anglophone Caribbean, institutional arrangements for tourism, tourism planning and development and tourism policy and decision-making. More recent research focuses on heritage tourism and customer experience management.

Eritha Huntley Lewis has over eighteen years' experience in adult education and training. Her areas of specialization are Tourism Policy and Planning, Events Management and Marketing. She has worked as Tourism Training Manager with the Tourism Product Development Company Ltd. (Jamaica) and has held several teaching positions at the University of the

West Indies (Mona campus), University of Technology Jamaica, Northern Caribbean University and University College of the Caribbean in Jamaica. Mrs. Huntley Lewis has a M.Sc. in Tourism and Hospitality Management from the University of the West Indies Mona and a Bachelor of Law with the University of London.

Acolla Lewis-Cameron is currently Dean of the Faculty of Social Sciences at the University of the West Indies (UWI), St. Augustine, Trinidad. Her educational achievements include a M.Sc. Hospitality and Tourism Education from the University of Surrey, UK and a Ph.D. in Tourism from Brunel University in the UK. Acolla's teaching experience in tourism planning and policy and marketing, and her research and consultancy in the above areas have provided an all-round understanding of the industry. She is the lead editor of text, '*Marketing Island Destinations: Concepts and Cases*' and co-author of the text '*Caribbean Tourism: Concepts and Cases*'. Acolla is actively engaged in research for both the private and public sectors in Trinidad and Tobago. She is passionate about empowering young people to develop to their full potential.

Twila-Mae Logan graduated from The Ohio State University with a Ph.D. in Business Administration (Finance and Economics) and a M.A. in Finance. She also obtained an M.B.A. from Dalhousie University and a B.Sc. in Mathematics and Meteorology from the University of the West Indies. She taught at the University of the West Indies, Nova Southeastern University and Messiah College at the undergraduate, masters and doctoral levels. She is currently an Associate Professor at the Chaplin School of Hospitality and Tourism Management—Florida International University. She has published papers related to the hospitality and tourism industries.

Kennedy Obombo Magio holds a Ph.D.. in Tourism Management from Universidad Autónoma de Occidente, Mexico. He is a distinguished member of the National System of Researchers in México (SNI 1) and his research interest is largely focused on tourism sustainability in the Mexican Caribbean. With over 30 publications, Kennedy has been awarded several scholarships and grants to contribute to tourism knowledge, including his current research fellowship and an Excellence Award by the Mexican Academy for Tourism Research (AMIT). He has undertaken a number of consultancies for public and private sector tourism organizations within Mexico and for international agencies.

Joanna Moses-Wothke A tourism professional with 15 years of tourism consultancy and academia experience in the Caribbean working on projects funded by major intergovernmental and governmental entities. She is currently the Dean of Academic Services at the TobagoHospitality and Tourism Institute, advocating for sound research writing, sustainable tourismplanning and policy development within the Caribbean. She volunteers as the CEO at an NGO responsible for the local coordination of the globally recognized Blue Flag and Green Key Programmes of the Foundation for Environmental Education. She served on the Board of Directors for governmental and civil society entities charged with the development and marketing of tourism in Tobago and provided expertise on human resource, research, product development and marketing matters.

Sherma Roberts is a Senior Lecturer in Tourism at the University of the West Indies, Cave Hill Campus. Having taught in the UK, she joined the Department of Management Studies in 2005, where she has worked hard to raise the profile of tourism education. Some of her accomplishments include hosting the first UWI wide international tourism conference, creating the opportunity for students in the OECS to pursue undergraduate tourism programmes at Cave Hill, developing a suite of postgraduate tourism programmes that are relevant to the region and collaborating closely with the local and regional industry for internships and guest lectures. In addition to teaching, she serves as Chair of the Tobago Tourism Agency and previously held positions as Chair of the Tourism Advisory Council in Barbados and Deputy Chair of the Barbados Tourism Product Authority. She has been instrumental in developing the participatory framework for the Tourism White Paper (2012–2021) in Barbados. Recently, she served as a panelist for the Caribbean Examinations Council developing the recently launched CAPE tourism syllabus and was Chief Examiner for the subject from 2014 to 2016. She has three co-edited books and recently published *Contemporary Caribbean Tourism: Concept and Cases*—which will be used by tourism students throughout the Caribbean archipelago and beyond. Sherma has written and presented papers in areas pertaining to community participation, corporate social responsibility, sustainable tourism, diaspora tourism, e-marketing and tourism entrepreneurship.

Clive Scott graduated from The Ohio State University with a Ph.D. in Agricultural Education. He holds a B.Sc. Degree in Animal Science and Agri-business from Tuskegee University and M.Sc. in Agribusiness Management from Alabama A&M University. He has held administrative positions at the College of Agriculture, Science and Education, The Ohio State University and Broward College. He taught at Nova Southeastern University for both graduate and undergraduate programmes. Dr. Scott is a Senior Research Fellow at the Mona School of Business and Management, University of the West Indies.

Ye Shen Ph.D. is an Assistant Professor in the Department of Experience Industry Management at California Polytechnic State University, San Luis Obispo. She completed her Ph.D. in Management at the University of Guelph in Canada and received her master's degree in Urban and Regional Planning at Peking University in China. She specializes in tourism experience design, destination marketing and technology applications in tourism and hospitality. Her research has been published in top-tier academic journals, such as *Tourism Management, Journal of Hospitality and Tourism Research* and *International Journal of Contemporary Hospitality Management.*

Shinelle Smith is a Tourism Lecturer at Tobago Hospitality and Tourism Institute. She is passionate about inspiring critical thinking in the field of education and challenges her students to use their knowledge to address real-life issues and to create viable solutions affecting tourism-oriented communities. Shinelle is also the founder of The Travel Taxi Ltd., a regional, tourism-based consultancy which promotes sustainable, eco-friendly travel and affords hospitality training to small and medium-sized businesses within the industry. Her company has consulted for properties in St. Kitts, Grenada, Dominica and Tobago. She holds a Bachelor of Science in Hospitality and Tourism, a Master of Science in Tourism Development and Management, a Master of Science in Innovation, Management and Entrepreneurship, a CELTA certificate from Cambridge and is currently pursuing qualifications in Education Technology and Education Administration. In her free time, her hobbies include travel, cooking and reading West Indian fiction.

David Smith has a Ph.D. in Zoology. He is currently the Coordinator of The Institute for Sustainable Development and Director for The Centre for Environmental Management at the University of the West Indies, Mona Campus. He has numerous years of experience in academia, the private sector, civil society and The United Nations focusing on topics such as Sustainability, Disaster Risk Reduction and Biodiversity Conservation. He has been involved in the successful implementation of several projects at the University that focus on Disaster Risk Management, hazard mapping, knowledge management, distance teaching and sustainable developmentresearch.

Kimberly Thomas-Francois, Ph.D., is a Post-Doctoral Research Fellow at Gordan S. Lang School of Business and Economics, School of Hospitality, Food and Tourism Management at the University of Guelph. She holds a Ph.D. in Management specialized in Service Management from the University of Guelph. She also holds a MSc. in Tourism and Hospitality Management and a B.Sc. in Management from the University of the West Indies. Kimberly has conducted research in multidisciplinary areas including agri-tourism linkages, customer engagement, and innovationin rural tourism. Currently in her post-doctoral research, she commenced a research programme in smart and virtual food retailing to investigate the digital food economy.

Aljoscha Wothke As a senior expert with broad experience in Caribbean SIDS, his main technical expertise is facilitating innovative stakeholder collaborations regarding sustainable green & blue economy, responsible tourism, protected area management, environmental resource management, marine and terrestrial conservation, and organizational capacity building. He is a founding Director of North East Tobago's most outstanding ENGO and has been project lead in award-winning interventions. He was the team leader for drafting the technical documents for NE Tobago's successful nomination as a UNESCO MaB Reserve, lead on drafting the dossier to list Tobago's Main Ridge Forest Reserve as the first Natural National Heritage Site, as well as the Sustainable Shark and Ray Management Plan for the MaB Reserve.

List of Figures

Fig. 2.1	Map of the Caribbean (*Source* http://www.turq.com/destmaps.html)	15
Fig. 3.1	Vulnerability-resilience nexus (*Source* Briguglio 2016)	47
Fig. 3.2	Thank you letter to Prime Minister Rowley from Prime Minister Mottley	50
Fig. 4.1	Cruise Terminals in Castries, St. Lucia	73
Fig. 4.2	The CT-MPA nexus in Soufriere	74
Fig. 4.3	Mixed Method Research Design	75
Fig. 5.1	Destination Sustainability Performance by Category	93
Fig. 8.1	The designated North-East Tobago UNESCO Man and the Biosphere (*Source* Developed by ERIC for the THA, 2019)	157
Fig. 8.2	Special Considerations for the UNESCO MaB Reserve Designation (*Source* Authors, 2021)	171
Fig. 10.1	Location of Negril and Runaway Bay in relation to other towns within Jamaica (*Source* Authors)	201
Fig. 10.2	Location of Crown Point and Speyside in relation to other towns within Tobago (*Source* Authors)	201
Fig. 10.3	Pictures showing an example of MSMTE's interviewed for this research Upper left: A glass bottom boat in Crown Point, Upper Right: Hotel offering accommodation in Speyside, Bottom Left: Craft hut in Speyside, Bottom Right: Craft enterprise in Runaway Bay. (*Source* Balkaran, T.)	202

Fig. 10.4 Resilient tourism businesses produced by the combination of measures from 4 main actors (*Source* Authors) 205
Fig. 11.1 Multiplier Effects 225
Fig. 12.1 Factors promoting the formation on the CTDA 251
Fig. 12.2 Views on sustainability 255
Fig. 12.3 Factors hindering entrepreneurial tourism development 257
Fig. 12.4 Main clientele for Castara 258
Fig. 12.5 Marketing strategies used 258
Fig. 12.6 Effectiveness of marketing tools 259
Fig. 12.7 Education levels among entrepreneurs in Castara 261
Fig. 14.1 Pillars of resilience (2018 *Source* CDEMA [, 4]) 291

List of Tables

Table 3.1	Shock events in the Caribbean 1995–2021	44
Table 4.1	Description of elite interviewees	74
Table 4.2	Perceived level of threat to the physical-ecological capacity in the CT-MPA nexus	81
Table 6.1	List of questions used for semi-structured interviews	121
Table 10.1	A breakdown of the numbers of interviews conducted at each site	204
Table 10.2	Suggested measures to promote resilience in the private sector	207
Table 10.3	Measures needed in the tourism industry to increase the resilience of MSME's	211
Table 11.1	Major source of agricultural purchases by size of property	231
Table 11.2	Purchase of meats from local farmers by size of property	232
Table 11.3	Problems purchasing agricultural produce from local farmer	233
Table 11.4	Problems purchasing agricultural produce from local farmers	234
Table 13.1	Destination recovery responses	276
Table 13.2	Sample of Caribbean countries COVID-19 tourism recovery responses	278

PART I

Background and Conceptual Context

CHAPTER 1

Introduction

Acolla Lewis-Cameron

Background

The operating environment for tourism continues to evolve in the increasingly crisis prone global tourism space. Destinations the world over are engaged in an ongoing exercise to effectively manage all components of the tourism system to ensure survival first and then increased competitiveness in the face of continuous shocks to the system. Most recently, the global tourism industry in 2020 has experienced unprecedented loss due to the COVID-19 pandemic unlike any other period of its growth and development. All major international tourism organizations are referring to 2020 as a period of unprecedented disruption as evidenced by massive decline in global tourism arrivals, export earnings from tourism and related employment. The United Nations World Tourism Organization (UNWTO) has designated 2020 the worst year in tourism history citing a 74% decline in international arrivals and an estimated loss of USD 1.3 trillion in export revenues; a loss that is 11 times greater than the fallout of the 2008–2009 global economic crisis (UNWTO 2021). A

A. Lewis-Cameron (✉)
Faculty of Social Sciences, University of the West Indies, St. Augustine, Trinidad
e-mail: acolla.lewis-cameron@sta.uwi.edu

A. Lewis-Cameron et al. (eds.), *Managing Crises in Tourism*,
https://doi.org/10.1007/978-3-030-80238-7_1

shock of this magnitude has the potential to redefine the operating environment for global tourism and brings to the fore the issue of the capacity of tourism systems to deal with such stressors by maintaining the stability of tourism-related economies.

Small Island Developing States (SIDS) such as the Caribbean islands by nature are highly vulnerable to external shocks. SIDS are often regarded as island countries and territories with a particular set of sustainable development challenges including small populations, limited resources, remoteness, susceptibility to natural disasters, vulnerability to external shocks, high transportation costs and fragile environments (UNEP 2014). These systemic challenges have been compounded by many SIDS' overdependence on one export or another. For the majority of the Caribbean, the overdependence has been on tourism which is viewed as a panacea for growth and development. On average, the tourism sector is said to account for approximately 30% of GDP for all SIDS and generates approximately $30 billion per year (Coke-Hamilton 2020). Their characteristics which propel them towards tourism for economic growth also jeopardize the sustainability of their tourism sectors which in turn is further exaggerated by tourism's inherent volatility (Hall 2012). This "developmental trap" presents a conundrum for the Caribbean as there are limited cost-effective opportunities to pivot away from dependency on tourism.

The concept of resilience, albeit highly contested, has therefore been a critical feature of Caribbean society given the high overdependence on tourism as a deliberate export strategy. Vulnerability and resilience studies in the context of tourism have been widely researched in general and specific to SIDS (Alberts and Baldacchino 2017; Economic Commission for Latin America and the Caribbean 2011; Moghal and O'Connell 2018; Scheyvens and Momsen 2008; Sheller 2020; Van Der Veeken et al. 2016; Weis et al. 2021). Within the SIDS context, particular reference is made to economic and environmental vulnerability with vulnerability being referred to as "the propensity of exposed elements such as physical or capital assets, as well as human beings and their livelihoods, to experience harm and suffer damage and loss when impacted by single or compound hazard events (Birkmann et al. 2013, 195). Resilience is regarded as the ability of an entity (society, community, region) to cope, absorb, withstand and recover from external and internal shocks (Haimes 2009; Hosseini et al. 2016; Simmie and Martin 2010).

An ECLAC (2011) report which sought to evaluate vulnerability and resilience of Caribbean SIDS under three dimensions (economic, environmental and social) discovered that across all dimensions, vulnerability and resilience differed based on several conditions including the degree of exposure to a particular shock and concludes that "….no single mechanism for building resilience to these impacts is applicable across the entire sub region" (p. 27). This was exemplified in Podhorodecka (2018) study which indicated:

> Some territories coped much better with the impact that the global financial crisis had on tourism. Tourism movement was best for the Seychelles, Bahamas, Maldives, and Saint Kitts and Nevis. The poorest results regarding tourist arrivals were for Antigua and Barbuda, Anguilla, Dominica, the British Virgin Islands, and Saint Vincent and the Grenadines. (p. 165)

Calgaro et al. (2014) assert that destination vulnerability and resilience is based on a number of elements such as cultural norms, power systems, values, political and economic ideologies, religious doctrines and agendas and expectations which combined is said to:

> …permeate the fabric of a destination and influence the nature and intensity of disruptive events, actions, reactions and consequences, and in turn, vulnerability and resilience. They are power-laden and deeply rooted in culture, history, religion and ideology. They shape governance structures and reinforce dominant ideologies, influence developmental decisions and destination characteristics, determine differential access to resources and their usage, influence business decisions and shape perceptions of risk and corresponding responses. Understanding these power dynamics and identifying the agendas that determine the form tourism activity takes in each destination, and tourist flows to those destinations, is therefore crucial to understanding and addressing destination vulnerability. (Calgaro et al. 2014, 353)

Concomitantly, the World Bank Group (2020, 18) asserts that "tourism resilience is a factor of both supply and demand-side variables". A study by Vítová et al. (2019) employing a gravity model framework with panel data for fifty-six SIDS discovered that the tourism infrastructure (supply components) played a key role in determining the level of tourism demand. Notwithstanding, it is argued that SIDS vulnerability is exaggerated and resilience diminished because of not only its dependency on tourism but also its development of tourism solely on demand factors and specifically, its dependency on particular types of tourism products, type of tourists and particular source markets (Calgaro et al. 2014).

The foregoing suggests that SIDS are vulnerable to a number of conditions determined by physical, social, economic and environmental factors or processes which increase their susceptibility to the impacts of hazards. However, the extent of their resilience or ability to absorb, accommodate, adapt to, transform and recover from the effects of a crisis in a timely and efficient manner is less well-known or documented in a single monograph. One of the earliest works on destination crisis management examines the effects of natural, health and political crises and how they are managed by a number of destinations. This book provides some valuable insights in terms of the details of the crisis and the public and private sector policy and marketing responses. Adding to the body of work on crisis and disaster management, Goeldner and Ritchie (2009) offer a more conceptual approach to this area of study including disaster classifications, prevention, planning and management frameworks and communication and recovery strategies. Writing specifically on tourism resilience, Lew and Cheer (2017) examine resilience and adaptation strategies used by destinations to combat environmental effects. The case studies are drawn from a number of destination contexts including islands, peripheral regions and developed or mature countries. In a similar vein, Cheer and Lew's (2018) *Tourism Resilience and Sustainability—Adapting to Social, Political and Economic Change* presents a strong analysis of the various resilience typologies and examines the intersectionality between resilience and sustainable tourism objectives. A number of cases with varying units of analyses (community, small tourism firms, islands, niche tourism expressions) are highlighted in this collection and examined for resilience outcomes.

Despite these very valuable contributions to the research area of vulnerabilities and resilience in tourism destinations, to date, no single monograph has examined the extent to which the juxtaposition of systemic vulnerabilities and the socio-psychological and institutional strengths of SIDS alluded to by Scheyvens and Momsen (2008) work to strengthen or undermine resilience efforts. Moreover, given that Caribbean SIDS have the unenviable nomenclature of being the most tourism-dependent region in the world, this book considers how can resilience strategies be effective in a binary system of tourism overdependence and island fragility, or even, whether resilience as conceptualized in the development literature has the same meaning for tourism-dependent Caribbean islands and territories. Pertinently, the chapters in this book ask the question "how is tourism resilience understood and practiced in Caribbean SIDS and what are the factors that inform, undermine or indeed redefine the sustainable resilience agenda for these territories?" It also presents strategic considerations to public and private sector practitioners in implementing measures to strengthen the competitive positioning of their destinations as they contend with the dynamism of the external and internal environments.

Organization of the Book

By spanning the geography of the Anglophone and Spanish Caribbean, this book is divided into five parts that offer a smorgasbord of conceptual and applied perspectives of tourism resilience in SIDS. Part I aptly provides the context for the book by firstly examining the features and characteristics of vulnerable Caribbean SIDS along with the unique challenges that arise due to isolation and exposure to increased risk and disasters. Small states are highly susceptible to a number of risks that make them more vulnerable than some of their larger counterparts. The risks range from economic to environmental, and if not well managed these can have a deleterious effect on the socio-economic fabric of many islands. The role that tourism plays in the growth and development of the islands is explored and the risks associated with an overdependence on the sector are also captured in this opening chapter. The issue of vulnerability must be understood in the context of resilience, which is the focus of the text. For island states where tourism is the lifeblood of the economies, the ability to adapt and to effectively manage change frames the nature and extent of resilience. Thus, the second conceptual chapter in Part I explores the concept of resilience and its application to SIDS and tourism. The discussion brings to light the many contestations around resilience understanding and its application in the context of destinations and businesses.

Parts II to IV of the book explore the vulnerabilities and resilience strategies adopted by specific Caribbean SIDS under the sub-headings of environmental resilience, social resilience and economic resilience, respectively. In discussing resilience in these territories, this section of the book takes the reader through the islands and captures the various ways in which communities have been able to surmount their vulnerabilities through communal resources, traditional beliefs and a variety of resilience strategies. In Part II, environmental resilience strategies were advanced for the territories of St. Lucia, Cozumel and Dominica, respectively. In Chapter 4, it was noted that the island of St. Lucia has benefitted consistently from cruise tourism over the last decade. However, the marine environment is approaching its physical-ecological carrying capacity limit. The author surmises that a system thinking approach and a shared commitment among stakeholders are critical interventions required at this time. Cozumel faces a similar challenge as its fragile ecosystem is threatened by ever-increasing tourist numbers. Both best practices and risk areas that are undermining Cozumel's environmental, social and economic sustainability are examined in this chapter. Whereas in the cases of St. Lucia and Cozumel the focus is on mitigation strategies to build resilience, in Chapter 6, there is a shift towards recovery strategies and the lessons to be learned in order to be more resilient. The case of Dominica and its tourism recovery strategies is reviewed in the aftermath of the devastation caused by Hurricane Maria.

Chapter 7 to nine captured in Part III place the spotlight on the people in the resilience discussion. In Chapter 7, the author recounts the devastation experienced by Hurricane Maria on the island of Puerto Rico. In this case, the author is particularly intrigued by the response of the adventure tourism community. The group of adventure tour operators anticipated the stressors to their businesses and came together as a community to recover from the damages caused by Hurricane Maria. North-East Tobago, a rural, underdeveloped and marginalized segment of the island is explored in Chapter 8. The area is socially and economically vulnerable to external shocks such as climate change, hurricanes and economic crises. An analysis of the path taken by North-East Tobago between 2015 and 2020 towards the sustainable management of its natural and cultural resources aimed at supporting cross-sectoral resilience to external shocks, while focusing on a responsible tourism approach with well-planned strategies to improve branding, marketing, competitiveness and diversification are elaborated. In Chapter 9, the author highlights the vulnerability of the youth in the face of the many economic challenges in Grenada. Strategies are put forward that create complements to current tourism products and services and at the same time identify opportunities for creative industries that can ensure sustainable continuity of life in times of disruptions.

Under the heading "Economic Resilience" in Part IV, the focus is on the ability of an economy to resist a particular shock and to recover rapidly to the previous level of growth or better. A core component of economic resilience is business resilience, which is the capacity for an enterprise to survive, adapt and grow in the face of turbulent change. In Chapter 10, the authors noted that disasters represent a growing risk for Micro, Small and Medium Enterprises (MSME's) which account for a significant proportion of businesses within the tourism industry in the Caribbean. A mixed methodology was used to investigate the factors influencing vulnerability to tropical cyclones and a number of measures were recommended to increase resilience within the private sector in Jamaica and Tobago. Another study was conducted on the island of Jamaica on the link between agriculture and tourism and the findings were presented in Chapter 11. The point is made that both sectors are highly vulnerable to external shocks and therefore can benefit from a partnership. The chapter makes recommendations for innovative ways in which the linkage between the hotel sub-sector and agricultural sector can be strengthened while also ensuring that local communities benefit from this linkage. In Chapter 12, the authors examine integrative entrepreneurship as a tourism resilience strategy in the rural community of Castara, Tobago. A critical analysis of the entrepreneurs and activities within the community as it relates to tourism-based collaborative entrepreneurship as a community reliance strategy is conducted. The authors further examine whether the collaborative approach can be deemed sustainable utilizing a participatory model which is largely reliant on the promotion of local entrepreneurship.

Part V of the book is appropriately titled "Contemporary Landscape and Reflections" as the closing chapters contemplate the Caribbean's response to the COVID-19 pandemic and reflect on the lessons learned about resilience experiences of Caribbean islands. In Chapter 13, the recovery strategies crafted by select Caribbean destinations in trying to balance lives and livelihoods during the COVID-19 pandemic are examined. This is followed by a discussion on the implications for tourism resilience in island destinations that are overdependent on tourism. In the Conclusion, the editors reflect on how tourism resilience is understood and practised in the Caribbean SIDS under study. Key lessons emerge as it is proposed that Caribbean island states progress towards the adoption of a transformative resilience agenda.

References

Alberts, A., and G. Baldacchino. 2017. Resilience and Tourism in Islands: Insights from the Caribbean. In *Tourism and Resilience*, ed. R.W. Butler, 150–162. Boston, MA: CABI.

Birkmann, J., O. Cardona, M. Pelling, and D. Alexander. 2013. Framing Vulnerability, Risk and Societal Responses: The MOVE Framework. *Natural Hazards* 67: 193–211.

Calgaro, E., K. Lloyd, and D. Dominey-Howes. 2014. From Vulnerability to Transformation: A Framework for Assessing the Vulnerability and Resilience of Tourism Destinations. *Journal of Sustainable Tourism* 22 (3): 341–360.

Cheer, J.M., and A.A. Lew. 2018. Understanding Tourism Resilience: Adapting to Social, Political, and Economic Change. In *Tourism, Resilience and Sustainability: Adapting to Social, Political and Economic Change*, ed. J.M. Cheer and A.A. Lew, 3–17. New York, NY: Routledge.

Coke-Hamilton, P. 2020. Impact of COVID-19 on Tourism in Small Island Developing States. *UNCTAD*, April 24, 2020. https://unctad.org/news/impact-covid-19-tourism-small-island-developing-states.

Economic Commission for Latin America and the Caribbean. 2011. *Study on the Vulnerability and Resilience of Caribbean Small Island Developing States (SIDS)*. Washington: ECLAC. https://www.cepal.org/sites/default/files/publication/files/38568/LCCARL354_en.pdf.

Goeldner, C.R., and J.R.B. Ritchie. 2009. *Tourism: Principles, Practices*. Philosophies: Wiley.

Haimes, Y.Y. 2009. On the Definition of Resilience in Systems. *Risk Analysis: An International Journal* 29 (4): 498–501.

Hall, C. Michael. 2012. Island, Islandness, Vulnerability and Resilience. *Tourism Recreation Research* 37 (2): 177–181.

Hosseini, S., K. Barker, and J.E. Ramirez-Marquez. 2016. A Review of Definitions and Measures of System Resilience. *Reliability Engineering & System Safety* 145: 47–61.

Lew, A., and J.M. Cheer, eds. 2017. *Tourism Resilience and Adaptation to Environmental Change: Definitions and Frameworks*. New York: Routledge.

Moghal, Z., and E. O'Connell. 2018. Multiple Stressors Impacting a Small Island Tourism Destination-Community: A Nested Vulnerability Assessment of Oistins, Barbados. *Tourism Management Perspectives* 26: 78–88.

Podhorodecka, K. 2018. Tourism Economies and Islands' Resilience to the Global Financial Crisis. *Island Studies Journal* 13 (2): 163–184.

Scheyvens, Regina, and Janet H. Momsen. 2008. Tourism and Poverty Reduction: Issues for Small Island States. *Tourism Geographies* 10 (1): 22–41.

Sheller, M. 2020. Reconstructing Tourism in the Caribbean: Connecting Pandemic Recovery, Climate Resilience and Sustainable Tourism Through Mobility Justice. *Journal of Sustainable Tourism*. https://doi.org/10.1080/09669582.2020.1791141.

Simmie, James, and Ron Martin. 2010. The Economic Resilience of Regions: Towards an Evolutionary Approach. *Cambridge Journal of Regions, Economy and Society* 3 (1): 27–43.

UNEP .2014. *Emerging Issues for Small Island Developing States*. Accessed January 20, 2021. https://sustainabledevelopment.un.org/content/documents/2173emerging%20issues%20of%20sids.pdf.

UNWTO. 2021. *2020: Worst Year in Tourism History with 1 Billion Fewer International Arrivals*. Accessed February 1, 2021. https://www.unwto.org/news/2020-worst-year-in-tourism-history-with-1-billion-fewer-international-arrivals#:~:text=Global%20tourism%20suffered%20its%20worst,World%20Tourism%20Organization%20(UNWTO).&text=The%20crisis%20has%20put%20between,small%20and%20medium%2Dsized%20enterprises.

Van Der Veeken, S., E. Calgaro, L. Munk Klint, A. Law, M. Jiang, T. De Lacy, and D. Dominey-Howes. 2016. Tourism Destinations' Vulnerability to Climate Change: Nature-Based Tourism in Vava'u, the Kingdom of Tonga. *Tourism and Hospitality Research* 16 (1): 50–71.

Vítová, P., J. Harmáček, and Z. Opršal. 2019. Determinants of Tourism Flows in Small Island Developing States (SIDS). *Island Studies Journal* 14 (2): 3–22.

Weis, K., C. Chambers, and P.J. Holladay. 2021. Social-Ecological Resilience and Community-Based Tourism in the Commonwealth of Dominica. *Tourism Geographies*. https://doi.org/10.1080/14616688.2021.1878267.

World Bank. 2020. *Rebuilding Tourism Competitiveness: Tourism Response, Recovery and Resilience to the COVID-19 Crisis*. Washington, DC: World Bank.

CHAPTER 2

Understanding Small Island States and Territories

Acolla Lewis-Cameron

Introduction

Small Island Developing States (SIDS) share features and experiences that set them apart from more developed mainland destinations. These island states are often regarded as island countries and territories with a particular set of sustainable development challenges including small populations, limited resources, remoteness, susceptibility to natural disasters, vulnerability to external shocks, high transportation costs and fragile environments (OCED 2018; UN-OHRLLS 2011). The distinct context of SIDS is captured in the differences in the role of tourism in the economy and the resources upon which the industry is dependent. In terms of the former, the tourism industry in many SIDS, particularly in the Caribbean, is the mainstay of the local economies with a significant degree of foreign ownership. With respect to the physical and human resources, the industry thrives on a fragile natural environment and because of the smallness of these islands, the environmental impacts are more severe than

A. Lewis-Cameron (✉)
Faculty of Social Sciences, University of the West Indies, St. Augustine, Trinidad
e-mail: acolla.lewis-cameron@sta.uwi.edu

A. Lewis-Cameron et al. (eds.), *Managing Crises in Tourism*,
https://doi.org/10.1007/978-3-030-80238-7_2

on mainland destinations. The purpose of this chapter is to examine the local context of SIDS by exploring these two main differences that set them apart from more developed mainland destinations and the unique challenges that arise owing to these peculiarities. The discussion proceeds firstly with a general insight into SIDS. An analysis of island tourism will follow focusing on its development, the role of tourism in the economy and the resources upon which the industry is dependent.

Islands: An Overview

Any analysis of island tourism must begin with what constitutes an island. Weaver (1995, 593) defines small island states or dependencies as "island or archipelagic states and dependencies with less than three million permanent residents and a land area of no more than 28,000 km squared". The majority of islands in the Caribbean, the area under study, fall within this definition with the exception of Haiti, Cuba and the Dominican Republic.

The Caribbean consists of several hundreds of islands grouped politically in some 30 island states. In a general sense, it has been considered as any country, region or island which is either in the Caribbean Sea or which touches upon it. Mackay and Spencer (2017, 46) define the Caribbean as "…large and small countries (in terms of both land mass and population); island states and countries on mainland continents; Spanish, Dutch, French and English-speaking countries; and independent and dependent territories". Figure 2.1 shows a diagrammatical representation of the Caribbean islands.

Economically, the realities of insularity and scale in the Caribbean islands, along with a broadly shared heritage of colonial status, have helped to foster a small island syndrome of underdevelopment and a defeatist "island state of mind" (DiPietro and Peterson 2017; Scheyvens and Momsen 2008). The majority of these islands rely upon a very limited range of economic activities and concomitant vulnerability to fluctuations in these sectors. The colonial period saw the introduction of and specialisation in a few plantation crops such as sugar, coffee, copra and bananas. Climate and soil conditions help to restrict the variety of agricultural products (Lockhart and Smith 1997). Therefore, many islands historically have relied on a few agricultural cash crops for foreign exchange.

Caribbean islands also suffer from restricted natural resource endowments, with few exceptions, notably Jamaica with bauxite and Trinidad

Fig. 2.1 Map of the Caribbean (*Source* http://www.turq.com/destmaps.html)

with petroleum. As a result, these SIDS are continuously struggling to identify activities which will contribute to the goals of long-term economic development. It is this combination of a limited array of economic activities and a restricted resource base that cause these islands to be heavily dependent on foreign aid infusions (Everest-Phillips 2014). As a corollary, the investment by external companies results in a high degree of foreign ownership of the local service industries which reinforces a dependency syndrome (Pratt 2015). This is exacerbated by the low levels of skilled and professional workers in islands resulting in the importation of labour thus perpetuating a situation of high unemployment among the locals. From an environmental standpoint, another notable characteristic of SIDS, as the United Nations (UN) (2010) highlighted, is the fragile natural ecology and its vulnerability to natural disasters like hurricanes.

The characteristics of SIDS in the Caribbean clearly indicate the challenges they face and signal the importance of the diversification of these economies. The UN (2010, 18) aptly describes their predicament by stating:

> SIDS are inherently economically vulnerable due to their remoteness and insularity, susceptibility to natural disasters, fragile ecology, limited institutional capacity, limited ability to diversify, strong dependence on a narrow range of exports, and high import content, particularly of strategic goods such as food and fuel, whose prices have exhibited high volatility.

Based on the need to break away from dependency on primary commodity exports and their fluctuating prices and the existence of only a few natural resources, many SIDS have sought to diversify their economies to tourism development. International mass tourism has thus become particularly important over the past four decades in these islands. Earlier perspectives by Weaver (1995) indicate that the persistent support for tourism development is derived from two main factors. First, governments perceive that most visitors are still attracted to islands' rich endowment of sun, sea and sand and by the resorts which these foster. Second, large luxury facilities equate with higher employment and gross revenues, while high levels of tourism development correlate positively with per capita incomes. However, current research indicates that governments are now increasingly concerned with enhancing the quality of life among its citizenry with employment of "pro-poor tourism" approaches and the overall pursuit of sustainable tourism development (Croes et al. 2018; Lee et al. 2015; Ridderstaat et al. 2016).

Tourism Development in the Caribbean

Small Island States (SIS) in the Caribbean have found it comparatively easy to attract tourists, and thus, the tourism industry has become the cornerstone of the majority of these island economies. Inherent characteristics such as distinct flora and fauna, year-round warm weather, picturesque beaches and unique cultural facets have made the Caribbean a region attractive for tourist consumption. Resultantly, the majority of islands have pursued tourism development as a panacea for economic development and have collectively earned the Caribbean the reputation of being the most tourism-dependent region in the world. Notwithstanding the devastating impacts of the COVID-19 virus on the Caribbean tourism sector from 2020, the region has registered tourism growth exceeding that of the global tourism industry. In 2019, the Caribbean experienced a 4.4% growth in arrivals outpacing the international growth rate of 3.8% (Caribbean Tourism Organization 2020). The World Travel and Tourism

Council (WTTC) (2020) noted that in 2019 the Caribbean among all other regions registered the largest travel and tourism total contribution to GDP and employment of 14.1% and 15.4%, respectively. Devastatingly, the region has registered a travel and tourism GDP contribution of -58% in 2020 demonstrating the highest level of economic loss from travel and tourism due to the global pandemic (WTTC 2021).

From a historical viewpoint, tourism became a significant factor in the economy of many islands in the 1960s. During colonial times, the function of the region was to supply a narrow range of staple exports of raw materials including sugar, coffee and other agricultural products to a limited number of metropolitan countries in exchange for essential food and manufactured imports. This plantation structure persisted even after the achievement of political self-determination by the majority of islands. However, high production costs, labour shortages, cane fires and the measures adopted by foreign governments to protect their own beet-sugar industry resulted in a new economic strategy moving the export base away from traditional agricultural staples toward other export sectors, such as light manufacturing and tourism (Holder 2013).

Many influential development theorists and policymakers contended that global trade, especially for primary commodities, was too erratic to form the principal "engine of growth" for Caribbean economies (Brohman 1996). Instead, it was believed that other outward-oriented strategies would offer a more secure and orderly basis for the generation of sustained growth. This view, which is popular among neo-liberals, is that developing countries like the Caribbean islands, at least during their initial stages of development, should uniformly specialise in primary exports rather than attempt to develop more sophisticated industrial sectors through state intervention that would not conform to comparative advantage based on factor proportions (Brohman 1996). Thus, the neoliberal strategy of outward-oriented development in the Caribbean islands centred on the promotion of new growth sectors such as tourism and other non-traditional exports such as manufacturing.

Governments and leaders of the Caribbean have certainly come to the view that tourism is anything from "an important", to "the most important", to "the only" means of economic survival for their states (Pattullo 1996; Cannonier and Burke 2019). However, in addition to the explosion in tourist arrivals to the islands, there are broader economic, socio-cultural and environmental issues related to the nature of the tourism industry that define tourism in the region and set it apart from tourism in mainland developed destinations. The two areas that would be examined in turn are the role of tourism in these island economies and the resources upon which they are dependent.

The Role of Tourism

Economic Dependence on Tourism

Tourism has been internationally recognised as providing tremendous economic benefits to any country that embarks upon its development. For several countries around the world, the tourism industry is highlighted as an income earner on the balance of payments. However, from the Caribbean perspective, Cannonier and Burke (2019, 103) write:

> Tourism remains a vital source of foreign exchange and a key component of economic activity in most of the small island Caribbean economies. In no other region of the world is the industry as vital to economic sustainability.

Tourism is more important in an economic sense to these island destinations than it is in mainland destinations. This is because it is invariably a larger and more significant part of the economies of these islands. Tourism in Germany represents in economic terms 9% of GDP, in the UK 10% and in Spain 12.4%. In the Caribbean, the share of tourism in GDP ranges from a third to a half or more than half for most islands. In Jamaica, the total contribution to GDP from tourism is 34.7%, while in Grenada it is 55.8% and in the Virgin Islands 73.1%. The United Nations World Tourism Organization (UNWTO) (2020) affirms that export revenues from tourism for island economies average around 9% exceeding the world average of 2% demonstrating the importance of tourism to these islands. Share of total exports from international tourism accounted in 2019 for 93% in St. Lucia, 70% in Antigua and Barbuda, 70% in Barbados and 54% in Jamaica. Travel and tourism are estimated to directly employ 413,000 persons in 2019 spanning across the accommodation, food and beverage, transport, travel trade and event sector. Of all the territories in the region, tourism means the most to the British Virgin Islands (BVI). It is responsible for generating 98.5% of GDP which is the highest share for any country in the world and accounts for 90% of direct and indirect employment (WTTC 2018).

Employment and income generation, increase in foreign exchange and tax earnings, attracting foreign investment and balancing the trade account are the main economic goals of tourism development for SIDS (Coke-Hamilton 2020; Pratt 2015; Seetanah 2011). For many of these islands, tourism is often either the only economic activity or one of two activities that sustain these economies. However, these islands that are heavily dependent on tourism experience a number of difficulties in generating ongoing employment for the locals and also in securing the much-needed foreign exchange.

Firstly, tourism creates direct employment in hotels, restaurants and beach clubs, and also generates a significant amount of indirect employment in the wholesale, retail and distributive trades. However, much employment is low paying and temporary or part time because of seasonal variations in tourist arrivals and offer little potential for development (Zampoukos and Ioannides 2011). Further to this, employment in the sector is skewed towards low-end jobs for the locals in the region. Stupart and Shipley (2013) noted that in Jamaica the tourism system was dominated by foreigners while Sealy (2018, 3–4) argues "Foreign domination of the hotel and distribution sectors of tourism has contributed to substantial leakage of foreign exchange revenues and contributed to social polarisation and social exclusion in the islands". Additionally, the perception of jobs in the industry is that they are menial, require low levels of qualifications, unsociable working hours and provide few prospects for advancement (Holder 2013; Sealy 2018). For these reasons, the region has chronic problems of attracting highly motivated and dynamic employees. Professions in more traditional fields such as medicine and law are still highly favoured by students, teachers and their parents. Gmelch and Gmelch (2012) discovered that in Barbados head teachers who possessed a limited understanding of career opportunities in the field of hospitality and tourism did not encourage their students to consider professions therein.

Secondly, the generation of income and increases in foreign exchange are heavily dependent on SIDS maintaining visitor numbers. However, the ability of island tourism planners to maximise the contribution of the tourism industry to the economy is severely constrained by a number of factors. Fluctuating visitor numbers as a result of global recessions or climatic variations, short lengths of stay and changing holiday fashions are problems that restrict planners. Because tourism is discretionary consumption expenditure, global economic downturns result in an especially steep decline in revenues for SIDS. The effects of such declines are further magnified for SIDS as tourists from metropolitan countries switch their vacations to closer, less expensive destinations during periods of financial hardships. Furthermore, an overreliance on single markets leaves industries extremely open to economic fluctuations in those islands and often creates a dependence on individual air links (Podhorodecka 2018). This is particularly problematic for islands in the Caribbean, which rely heavily on international carriers.

The core problem with SIDS being economically dependent on tourism lies in the fact that,

> The collapse in tourist arrivals not only directly affects income and employment in airlines, ground transport and hotels, but also adversely affects the rest of the economy, including agriculture and construction. Falling tourism, and subsequently, reducing tax revenues, will exacerbate fiscal balances of many small island economies and also reduce the flow of foreign direct investment (FDI), as the tourism sector is typically the largest recipient of FDI (United Nations Department of Economic and Social Affairs 2020, 3)

Foreign Dominance of the Tourism Industry

In SIDS where tourism is the leading economic sector, foreign ownership, high leakage and expatriate domination of management are at high levels (Nunkoo et al. 2010; Hampton and Jeyacheya 2020). Guam, Fiji, the Cook Islands and the Caribbean exemplify this. This foreign dominance of the tourism industry is closely linked to the initial rapid industrialisation period of these SIDS. The financial assistance from the International Monetary Fund (IMF) coupled with the direct investment by transnational corporations has played an integral role in the development of the tourism industry in SIDS. With low levels of income, scarce resources, infrastructural constraints in terms of the provision of roads, hotels and tourist-promotion programmes, and high operational costs, governments in SIDS provided generous incentives to potential investors from developed countries to pursue tourism development. These incentives usually consisted of a variety of tax-free concessions including the right to import duty-free materials and start-up equipment for hotels, exemption from land tax and capital levies, tax holidays, sometimes lasting up to 35 years, and the repatriation of investment and profits (Ambrosie 2015; Pattullo 1996).

The IMF has included tourism as part of its Structural Adjustment Programmes (SAPs). The SAPs, which are preconditions for the approval of financial assistance, required the indebted country to be integrated into the global economy; deregulate and liberalise its economy and; shift from an agriculture-based to a service industry-based economy (Chavez, 1999). The SAP opens up the local economy to foreign investments and transnational corporations, while eliminating subsidies and protection to local industries. The inevitable acceptance of the investment from these transnationals virtually means effectively transferring the control of one's tourism industry to them with marginal benefits for the local populace. Funding from the World Bank has also been available to Caribbean states for tourism project and today continues to perpetuate dependency on foreign investment (Williams 2012).

Transnationals have capitalised on the continuous growth in the global tourism industry and the opening up of the majority of indebted small island economies. The General Agreement on Trade in Services (GATS), a legal and operational framework for the gradual elimination of barriers to international trade in services, is encouraging the dominance of the transnationals in SIDS. In short, GATS makes it easier for these corporations to invest in the local tourism industries of SIDS. Among others, it removes restrictions on foreign corporations' abilities to transfer staff from one country to another and enables them to use trademarks, create and operate branch offices abroad, and more importantly, to repatriate their earnings to their mother companies abroad (Chavez 1999).

Airlines and tour operators are the two key players in this process. The competitive advantage of the tour wholesalers lies in their doubly strategic position between all principal suppliers and consumers. Their power derives from the enormous volumes they can command, their pivotal familiarity with diverse market segments and the capacity to shift tourist flows from one destination to another (Picazo and Moreno-Gil 2018). Package holidays in SIDS are supplied and marketed mainly by firms from industrialised countries. Moreover, payments for holiday expenditure are often made to travel agents and tour operators in tourist origin countries so that only a percentage reaches the destination.

Because of the monopolistic controls exerted by transnationals over the ownership and organisational structure of the tourism sector, the tourism industry in SIDS replicates problems of dependency, internal disarticulation and foreign exchange leakages. With the transnational corporations having such an immense stake in the industry in SIDS, Zappino (2005, 14) suggests,

> In the Caribbean, such as in many other developing countries, the powerful trans-national corporations (TNCs) continue to dominate the international tourism market (estimates suggest that about 80% of international mass tourism is controlled by TNCs). These companies have an almost unhindered access to markets and use this to drive down the cost of supplies. The result is a limited level of revenue retention in the destination or host countries. It has been estimated that in developing countries, on average, at least 55% of tourism expenditure flows back out of the destination country, while in the Commonwealth Caribbean it rises to 75%.

The foreign dominance of the industry in SIDS raises the issue of power and control. The concern for small island states is that the locus of control over the tourism development process is shifting from the people that are most affected by development, the host community, to the tourist-generating regions. Many authors argue that tourism is a conduit for continued colonialism of SIDS which allows for continued domination and perpetuation of a dependency syndrome (Burns 2008; Holder 2013; Wong 2015). Foreign domination and external dependency seriously reduce tourism's potential for generating broad based growth, as well as the net financial advantages that the industry brings to SIDS. Dehoorne et al. (2011) argue "...development mechanisms...in favor of furthering the plantation cycle: externally dependent economies, control of the marketplace by transnational groups and heightened competition with the production of a sole product".

A further negative impact of foreign domination of the tourism industry in SIDS has been the loss of control over local resources, which adversely affects the socio-economic and ecological wellbeing of the host communities. This is closely tied to the lack of planning on the part of many governments in SIDS. Peterson (2020) notes that currently in Aruba even after five decades of tourism plans and policies local people still find themselves in a situation where decisions governing their lives, even those that address local interests, are normally made elsewhere according to the narrow interests of those that control the tourism industry. Peterson (2020) states "...the realization of these master plans has, nevertheless, resulted in a system of tourism specialization, supply, and style that is disconnected from society and the local community". The need to satisfy and to protect the investment of the transnationals meant that these corporations received preferential treatment over the years with respect to the exploitation of the local resources. This is evident in the distribution of land in SIDS. Pattullo (1996) observed that the growth of tourism affected both the availability and the price of land, not only putting it out of the reach of locals, but also reducing the pool of land for agriculture and other uses. Sealy (2018) describes the situation in Barbados, Cuba and Jamaica where access to beaches by locals have been prohibited due to privatisation of such lands by foreign entities. She also speaks to the use of land for golf courses which many locals are not afforded the benefit of membership thereafter or are restricted by high membership fees.

Foreign domination of the tourism industry also contributes to the overseas leakage of a substantial portion of the earnings of SIDS in the Caribbean. In most of the Caribbean, the level of leakages averages around 70%, which means that for every dollar earned in foreign exchange, 70% is lost in imports. In the Bahamas leakages can be as high as 90% (Pattullo 1996). A 2016 study on Jamaican hotels indicated that hotels imported one-third of their food and fixtures amounting to an annual business leakage for the local manufacturing sector of $65.4 billion dollars (Jamaica Gleaner 2016). Foreign exchange earnings are also lost by way of expatriate staff salaries, profit repatriation of transnationals and rising consumption by locals of imported goods made available through the industry. Problems of excessive foreign exchange leakage within tourism have been aggravated by the sector's lack of articulation with other parts of the local economy, especially agriculture. Food and most finished goods required by hotels to satisfy tourist demand are brought largely from the outside, more specifically from metropolitan countries, which requires a considerable amount of foreign exchange. Only in islands like Jamaica and Trinidad can it be reasonably claimed that local products are predominant in the industry. Poor linkages in the islands can restrict the circulation of revenue through the domestic economy, producing lower multiplier effects in terms of both employment and income for the local population. Brohman (1996, 56) commented that generally, lower multipliers have been associated with highly concentrated, large-scale, foreign-owned corporations, while higher multipliers have been connected to more dispersed, smaller-scale, locally owned operations that tend to be better linked to the local economy.

Tourism's Resources

The Caribbean Tourism Product

The traditional consumer-perceived generic product of SIDS in the Caribbean is crystallised in the phrase, "escape the winter to tropical sun, sea and sand". The Caribbean islands are in a general sense tourist receiving destinations supplying primarily hospitality-based services including accommodation and catering facilities, infrastructural and other destination-specific facilities. Mainland destinations in many cases provide the airlift between metropolitan destinations and the islands and the vacation packages are assembled and paid for in the originating country. Thus, the tourism product in the Caribbean is hospitality based with a significant emphasis on accommodation services (McBain 2007).

The tourism industry in SIDS depends on a delicate mix of a fragile natural environment and the hospitality of the locals. The natural attractions of these islands form the heart of the tourism resource base. They include year-round pleasant climates, relatively unspoiled beaches and reefs, undisturbed and preserved natural environments such as wildlife, flora and fauna and mountains. These features and more are evident in the Caribbean islands ranging from the lush mountains of Jamaica to the tropical rainforests of Dominica to the historic plantation houses of Nevis. The impact of tourism on these resources in SIS has far greater implications than in the more developed mainland destinations.

A Fragile Natural Environment

Tourism in the Caribbean relies heavily on the region's environmental resources for economic growth and development. The natural environment is a complex set of marine and land-based ecosystems that are interlinked in such a way that when one is affected so are the others. Island ecosystems, such as those of the Pacific and the Caribbean, are substantially different from those of continental areas. They are characterised by limitations of space but has a larger number of unique flora and fauna and are also described as a "global marine diversity hotspot"

(Miloslavich et al. 2010). Thus, it is argued that SIDS can be significantly damaged for generations and perhaps forever by unplanned and uncontrolled tourism.

Barnett and Waters (2016) note that the small scale of an island's physical resources causes it to be far more vulnerable to exogenous shocks and natural disasters and is severely affected by over development. Tourism in the Caribbean is heavily resource based and is traditionally crafted around the traditional sun-sea-sand type tourism. However, environmental pressures threaten this key selling point on which Caribbean tourism depends. In particular, the last two decades has brought with it a sobering reality of climate change and the resultant effects on small islands and the tourism-dependent resources. In the Caribbean, increased temperatures, warmer seas and rising sea levels have contributed to the reduction or loss of beaches, coral bleaching, loss of mangroves and salt marshes and flooding (Mackay and Spencer 2017). Also, intensification of storms has caused mass destruction as recent as 2017 and 2019 in the Caribbean with Hurricane Maria and Dorian, respectively, obliterating whole islands and all economic activity. The island of Dominica incurred damages and losses equalling to 225% of GDP from the passing of Hurricane Maria.

Additionally, the Caribbean environment also faces threats from the improper disposal of waste generated by tourism activities. Over the years, there have been national and regional discussions over the improper disposal of untreated liquid waste by hotels and resorts located along Caribbean coastlines. De Cuba et al. (2008) in their study of waste-to-energy systems in the Caribbean indicate that tourists to the Caribbean generate twice as much solid and liquid waste per capita compared to local residents. They drew reference to the installation of 150 small sewage treatment systems in hotels in Trinidad and Tobago but noted that many of these systems were poorly maintained and provide limited benefits. Furthermore, Carrier and Macleod (2005) indicated that improper disposal of human waste and sewage from tourism activities in Montego Bay, Jamaica caused degradation of the water quality which adversely affected the fishing stock in the area thereby reducing income for fisherfolk in the area.

The simple presence of tourists can have adverse environmental impacts in some particularly sensitive ecological systems. In the Caribbean, a common example is the impact on coral reefs from various cruise ships and recreational users. In the Cayman Islands, 15 acres of coral reefs in under threat by plans to build two cruise ship docks. Roberts (2019) writes "Environmental campaigners warn the George Town Harbour project will see 22 acres of the seabed dredged, and silt sedimentation will turn the "crystal-clear aquamarine waters to murky white". Even with dissenting views, oftentimes projects of similar nature forge ahead due to an appreciation of economic gains rather than environmental preservation. In addition, the cruise liners have been deemed culprits in the illegal dumping of waste in Caribbean waters contributing to the demise of coral reefs. In 2019, Carnival Corporations paid a criminal penalty of US $20 million for releasing food and plastic waste into the waters of the Bahamas (Nace 2019) while in 2015 divers captured video footage of the anchor of a cruise ship dragging across the coral reef in Grand Cayman turning the coral into dust (Pursell 2015). Moscovici (2017, 368) state:

> Of the 109 countries globally with coral reefs, almost half of them have seen some damage from cruise ship anchors, sewage dumping, tourists breaking off chunks of coral, and commercial harvesting for sale to tourists... an anchor dropped in a coral reef for a day, impacted an area about half the size of a football field, most of which died. It could take upward of fifty years for coral to recover, if at all.

In the Caribbean, tourist facilities are often built in environmentally sensitive areas. The majority of tourism facilities are located within 800 m of the high-water mark, and most tourist activity takes place in the area between the back bays and fronting reefs. Large concrete hotels have been built close to the high-water mark in Jamaica, marinas for yachts in Trinidad and deep-water harbours for cruise ships in Tobago have been constructed. An extreme case is noted in Jamaica where Montego Bay's international airport was built on a wetland, while at nearby Ocho Rios, 40 acres of swamp were turned into a resort with 4,000 beds and a cruise ship pier (Pattullo 1996).

In research conducted by the CTO (2020), they identified three key factors that can contribute to the harmful impact of tourism on the environment. First, in many cases there is a lack of awareness on the part of some of those making decisions about tourism development of the social, economic and environmental balance to be pursued in achieving sustainable development. Second, weak institutional frameworks with inadequate controls lead to tourism development which is both inappropriate and intrusive and often capable of causing irreversible damage to the natural environment. Third, "unfairly traded tourism" which precludes local communities from sharing in the benefits of the industry. From more recent research, Peterson (2020) sums it up by stating that "…the fundamental lack of tourism governance and policy prudence by public and private stakeholders, in addition to excluding civic society and local communities from tourism decisions and development, have fostered the unrestricted expansion and negative externalities of tourism growth".

The catalogue of environmental destruction directly attributed to the growth of the tourist industry in the Caribbean is long including the erosion of beaches, the breakdown of coral reefs and the dumping of waste. The natural environment upon which tourism depends is fragile and is severely threatened by increased tourist activity. Therefore, tourism development in the Caribbean needs to be carefully planned because these islands simply do not have the depth of resources unlike continental areas to allow for recovery from any major environmental damage.

Conclusion

Tourism is the lifeblood of these small island economies. For the locals, they are either directly or indirectly dependent on earnings from the tourism industry. Thus, any downturns in the industry send a rippling effect throughout the economy. Both the human and the natural environments are the resources upon which the industry thrives. Therefore, the sustainability of the environment and the attitude of the residents towards tourism development and tourists are crucial for the industry's success. The foreign element dominates the industry both in terms of the management of the resources and the flow of tourists from the metropolitan countries to the islands. The result of this has been a form of dependency development that has implications for the long-term sustainability of the industry. This chapter has examined the unique challenges that arise for Caribbean SIDS due to isolation and exposure to increased risk and disasters. These challenges reveal the extent to which SIDS are vulnerable to the various external shocks. Understanding the challenges that face many SIDS is particularly important. Equally significant is an understanding of the strengths of these islands and the ways in which these strengths can be leveraged to manage the crises and the range of external shocks. The following chapter provides a foundational understanding of the scope of tourism resilience in SIDS from an environmental and socio-economic perspective.

References

Ambrosie, L.M. 2015. *Sun & Sea Tourism: Fantasy and Finance of the All-Inclusive Industry*. Lomdon: Cambridge Scholars Publishing.

Barnett, J., and E. Walters. 2016. Rethinking the Vulnerability of Small Island States: Climate Change and Development in the Pacific Islands. In *The Palgrave Handbook of International Development*, ed. Jean Grugel and Daniel Hammett, 731–748. London: Palgrave Macmillan.

Brohman, J. 1996. New Directions in Tourism for Third World Development. *Annals of Tourism Research* 23 (1): 48–70.

Burns, P. 2008. Tourism, Political Discourse, and Post-colonialism, Tourism and Hospitality. *Planning & Development* 5 (1): 61–71.

Cannonier, C., and M.G. Burke. 2019. The Economic Growth Impact of Tourism in Small Island Developing States—Evidence from the Caribbean. *Tourism Economics* 25 (1): 85–108.

Caribbean Tourism Organisation. 2020. *Record Year for Caribbean Tourism*. Accessed April 1, 2021. https://www.onecaribbean.org/statistics/annual-reviewsprospects/#:~:text=Record%20Year%20for%20Caribbean%20Tourism&text=Stayover%20arrivals%20grew%20by%204.4,growth%20rate%20in%20the%20Americas.

Carrier, J.G., and D.V. Macleod. 2005. Bursting the Bubble: The Socio-Cultural Context of Ecotourism. *Journal of the Royal Anthropological Institute* 11 (2): 315–334.

Chavez, R. 1999. Globalisation and Tourism: Deadly Mix for Indigenous Peoples. *Third World Resurgence* 103: 1–6.

Coke-Hamilton, P. 2020. *Impact of COVID-19 on Tourism in Small Island Developing States*. Accessed April 3, 2021. https://unctad.org/news/impact-covid-19-tourism-small-island-developing-states.

Croes, R., J. Ridderstaat, and M. van Niekerk. 2018. Connecting Quality of Life, Tourism Specialization, and Economic Growth in Small Island Destinations: The Case of Malta. *Tourism Management* 65: 212–223.

De Cuba, K., F. Burgos, R. Contreras-Lisperguer, and R. Penny. 2008. *Limits and Potential of Waste-to-Energy Systems in the Caribbean. Department of Sustainable Development: Organization of American States*. Accessed February 20, 2021. https://www.oas.org/dsd/reia/Documents/LimitsandPotentialofWastetoEnergy.pdf.

Dehoorne, O., C. Murat, and N. Petit-Charles. 2011. International Tourism in the Caribbean Area: Current Status and Future Prospects. *Études caribéennes* (16). https://doi.org/10.4000/etudescaribeennes.4713.

DiPietro, R.B., and R. Peterson. 2017. Exploring Cruise Experiences, Satisfaction, and Loyalty: The Case of Aruba as a Small-Island Tourism Economy. *International Journal of Hospitality & Tourism Administration* 18 (1): 41–60.

Everest-Phillips, M. 2014. *Small, So Simple?: Complexity in Small Island Developing States*. Singapore: UNDP Global Centre for Public Service Excellence.

Gmelch, S., and George Gmelch. 2012. *The Parish Behind God's Back: The Changing Culture of Rural Barbados*, 2nd ed. Long Grove, IL: Waveland Press.

Hampton, M.P., and J. Jeyacheya. 2020. Tourism-Dependent Small Islands, Inclusive Growth, and the Blue Economy. *One Earth* 2 (1): 8–10.

Holder, J. 2013. *Caribbean Tourism*. Jamaica: Canoe Pres.

Jamaica Gleaner. 2016. *Hotel Import Bill Estimated at $70b*. Accessed Februay 4, 2021. https://jamaica-gleaner.com/article/business/20160202/hotel-import-bill-estimated-70b.

Lee, Donna, M. Hampton, and J. Jeyacheya. 2015. The Political Economy of Precarious Work in the Tourism Industry in Small Island Developing States. *Review of International Political Economy* 22 (1): 194–223. https://doi.org/10.1080/09692290.2014.887590.

Lockhart, D.G., and D. Smith, eds. 1997. *Island Tourism: Trends and Prospects*. London: Cassell.

Mackay, E.A., and A. Spencer. 2017. The Future of Caribbean Tourism: Competition and Climate Change Implications. *Worldwide Hospitality and Tourism Themes* 9 (1): 44–59. https://doi.org/10.1108/WHATT-11-2016-0069.

Mc Bain, H. 2007. *Caribbean Tourism and Agriculture: Linking to Enhance Development and Competitiveness*. Port of Spain: United Nations, ECLAC.

Miloslavich P., J.M. Díaz, E. Klein, J.J. Alvarado, C. Díaz, and J. Gobin. 2010. Marine Biodiversity in the Caribbean: Regional Estimates and Distribution Patterns. *PLoS ONE* 5 (8): e11916. https://doi.org/10.1371/journal.pone.0011916.

Moscovici, D. 2017. Environmental Impacts of Cruise Ships on Island Nations. *Peace Review* 29 (3): 366–373. https://doi.org/10.1080/10402659.2017.1344580.

Nace, T. 2019. *Carnival Cruise to Pay $20 Million After Admitting to Dumping Plastic Waste in the Bahamas*. Accessed April 1, 2021. https://www.forbes.com/sites/trevornace/2019/06/11/carnival-cruise-to-pay-20-million-after-admitting-to-dumping-plastic-waste-in-the-bahamas/?sh=5a2db4fe3a6f.

Nunkoo, R., D. Gursoy, and T.D. Juwaheer. 2010. Island Residents Identities and Their Support for Tourism: An Integration of Two Theories. *Journal of Sustainable Tourism* 18 (5): 675–693.

OECD. 2018. *Making Development Co-operation Work for Small Island Developing States*. Paris: OECD Publishing.

Office of the High Representative for the Least Developed Countries, Landlocked Developing Countries and Small Island Developing States (UN-OHRLLS). 2011. *Small Island Developing States Small Islands Big(Ger) Stakes*. Accessed December 11, 2019. http://unohrlls.org/custom-content/uploads/2013/08/SIDS-Small-Islands-Bigger-Stakes.pdf.

Pattullo, P. 1996. *Last Resorts: The Cost of Tourism in the Caribbean*. London: Cassell.

Peterson, R.R. 2020. Over the Caribbean Top: Community Well-Being and Over-Tourism in Small Island Tourism Economies. *International Journal of Community Well-Being*, 1–38. https://doi.org/10.1007/s42413-020-00094-3.

Picazo, P., and S. Moreno-Gil. 2018. Tour Operators' Marketing Strategies and Their Impact on Prices of Sun and Beach Package Holidays. *Journal of Hospitality and Tourism Management* 35: 17–28.

Podhorodecka, K. 2018. Tourism Economies and Islands' Resilience to the Global Financial Crisis. *Island Studies Journal* 13 (2): 163–184.

Pratt, Stephen. 2015. The Economic Impact of Tourism in SIDS. *Annals of Tourism Research* 52: 148–160.

Pursell, R. 2015. *Video Captures Cruise Ship Destroying Protected Coral Reef*. Accessed February 6, 2021. https://www.mensjournal.com/adventure/video-of-cruise-ship-destroying-coral-reef/.

Ridderstaat, J., R. Croes, and P. Nijkamp. 2016. The Tourism Development–Quality of Life Nexus in a Small Island Destination. *Journal of Travel Research* 55 (1): 79–94.

Roberts, L. 2019. *Caribbean Coral Reef Could Be Destroyed to Make Way for Cruise Ships in the Cayman Islands*. Accessed April 20, 2021. https://www.telegraph.co.uk/news/2019/10/11/caribbean-coral-reef-could-destroyed-make-way-cruise-ships-cayman/.

Scheyvens, R., and J.H. Momsen. 2008. Tourism and Poverty Reduction: Issues for Small Island States. *Tourism Geographies* 10 (1): 22–41.

Sealy, Wendy. 2018. From Colonialism to Transnationalism: The Neo-colonial Structure of Caribbean Tourism. *Journal on Tourism & Sustainability* 1 (2): 81–92.

Seetanah, B. 2011. Assessing the Dynamic Economic Impact of Tourism for Island Economies. *Annals of Tourism Research* 38 (1): 291–308.

Stupart, C.A., and R. Shipley. 2013. Jamaica's Tourism: Sun, Sea and Sand to Cultural Heritage. *Journal of Tourism Insights* 3 (1): 4.

United Nations. 2010. *Trends in Sustainable Development Small Island Developing States*. Accessed December 12, 2020. https://www.cbd.int/islands/doc/sids-trends-report-v4-en.pdf.

United Nations Department of Economic and Social Affairs. 2020. *The COVID-19 Pandemic Puts Small Island Developing Economies in Dire Straits*. Accessed April 17, 2021. https://www.un.org/development/desa/dpad/wpcontent/uploads/sites/45/publication/PB_64.pdf.

United Nations World Tourism Organization. 2020. *Tourism in SIDS: The Challenge of Sustaining Livelihoods in Time of COVID-19*. Accessed April 20, 2021. https://www.e-unwto.org/doi/pdf/10.18111/9789284421916.

Weaver, D.B. 1995. Alternative Tourism in Montserrat. *Tourism Management* 16 (8): 593–604.

Williams, T.R. 2012. Tourism as a Neo-Colonial Phenomenon: Examining the Works of Pattullo & Mullings. *Caribbean Quilt* 2: 191.

Wong, A. 2015. Caribbean Island Tourism: Pathway to Continued Colonial Servitude. Etudes Caribéennes 31–32. https://doi.org/10.4000/etudescaribeennes.7524.

World Travel and Tourism Council. 2018. *Caribbean Resilience and Recovery: Minimising the Impact of the 2017 Hurricane Season on the Caribbean's Tourism Sector*. Accessed February 19, 2021. https://wttc.org/Portals/0/Documents/Reports/2018/Caribbean%20Recovery%20Report%20-%20Full%20Report%20-%20Apr%202018.pdf?ver=2021-02-25-182520-540.

World Travel and Tourism Council. 2020. *Regional Overview 2020*. Accessed January 24, 2021. https://wttc.org/Portals/0/Images/EIR2020-Infographic-Map-Regional.png?ver=2021-04-06-170812-653.

World Travel and Toruism Council. 2021. WTTC Research Reveals Travel & Tourism Sector's Contribution to Caribbean GDP dropped by $33.9 billion in 2020. Accessed May 30, 2021. https://wttc.org/News-Article/WTTC-research-reveals-Travel-Tourism-sectors-contribution-to-Caribbean-GDP-dropped-by-34-billion-in-2020.

Zampoukos, K., and D. Ioannides. 2011. The Tourism Labour Conundrum: Agenda for New Research in the Geography of Hospitality Workers. *Hospitality & Society* 1 (1): 25–45.

Zappino, Vincenzo. 2005. *Caribbean Tourism and Development: An Overview* [Final Report]. Accessed October 10, 2019. http://ecdpm.org/wpcontent/uploads/2013/11/DP-65-Caribbean-Tourism-Industry-Development2005.pdf.

CHAPTER 3

Conceptualizing Resilience in Small Island States

Sherma Roberts

Introduction

While the concept of resilience arguably has its natal roots in engineering, it is most widely associated with the ecological sciences and the seminal work of Holling (1973, 14) who defined resilience as 'the measure of persistence of systems and of their ability to absorb change and disturbance and still maintain the same relationships between populations or state variables.' Conceptualized in this way, resilience involves the ecosystem's inherent ability to recover from various shocks and return to a state of equilibrium. Unsurprisingly, social scientists have reached into the natural sciences and coopted the term for application into public policy and community discourses, seeing it as a relevant analytical framework, but also critiquing its occlusion of the social dimension, given the centrality of human actors in shaping systems and processes (Hall 2018). From an international policy perspective, resilience is seen as 'the ability

S. Roberts (✉)
Faculty of Social Sciences, Department of Management Studies, University of the West Indies Cave Hill Campus University Drive, St. Michael, Barbados
e-mail: sherma.roberts@cavehill.uwi.edu

A. Lewis-Cameron et al. (eds.), *Managing Crises in Tourism*,
https://doi.org/10.1007/978-3-030-80238-7_3

of a system, community or society exposed to hazards to resist, absorb, accommodate to and recover from the effects of a hazard in a timely and efficient manner, including through the preservation and restoration of its essential basic structures and functions' (United Nations Office for Disaster Reduction 2012). Approaches to understanding resilience now therefore include engineering resilience, ecological resilience, socio-ecological resilience, community resilience and evolutionary resilience (Cutter et al. 2008; Folke 2006; Sgrò et al. 2010).

The complexity, heterogeneity and susceptibility of the tourism industry to endogenous and exogenous shocks have meant that resilience as an organizing concept has been embraced by the industry and related public and private sector organizations. Within tourism, resilience is generally understood to be 'the ability of a destination (community, ecosystem, resources) to maintain its quality appeal, as well as its traditional way of life and natural processes, in the face of the effects of tourism and related developments' (Butler 2018, 116). Among tourism-dependent Caribbean small island states, the pertinence of resilience as a sustainability mechanism is accepted a priori, largely as a consequence of these islands' tourism overdependence and natural disaster vulnerabilities (CDEMA 2009; CDEMA and CARICOM 2018). While embracing the UNISDR resilience definition which is underpinned by the 'build back better' engineering resilience philosophy, the Caribbean Disaster Management Agency (CDEMA) has also adopted a 'bounce forward quickly' resilience perspective. This approach which is the one that is employed in this chapter recognizes the high dependence on tourism for foreign exchange revenues and employment, and therefore, the urgency of the vast majority of Caribbean island states to regain normalcy *and* effect transformation, where needed. The CDEMA perspective also acknowledges that within the region there are systems and power relationships which undermine Caribbean development, exposing it to a number of vulnerabilities, and these must be challenged and new pathways found if the region is to be truly resilient (CDEMA and CARICOM 2018).

The aim of this chapter therefore is to broadly examine conceptualizations of resilience, its determinants and critiques, and the ways and extent to which resilience ideas have been integrated into SIDS and tourism resilience research and practice. Tourism resilience as the analytical framework of this book would also be briefly discussed. The chapter also raises questions about the binary reality of tourism's inherent vulnerability and SIDS' tourism overdependence, and whether resilience is at all possible, if as a region the Caribbean is unable or afraid to confront transformation as a critical outcome of resilience.

Unpacking Resilience—Meanings, Tensions, Critiques

At the recently concluded online conference themed *Researching Resilience in Islands*, presenters highlighted the many contestations around the word resilience which have developed as a result of disciplinary positioning, cultural factors or policy cooptation (Foley 2020). Holling (1973) was the first to conceptualize resilience and did so from an ecological systems perspective, where resilience was construed as the ability to bounce back or regain some level of equilibrium from shocks. In this regard, resilience did not lead to transformation or change, but to maintaining the existing structure (Mitchell and Harris 2012). Contemporary ecological perspective scholars have reconciled that while the system has the ability to anticipate, absorb, accommodate or recover from a hazardous event, resilience can also lead to transformation and ensuring longer term development, despite the effects of climate change and other anthropogenic behaviors (Cutter et al. 2008; Dodman et al. 2008; Folke 2006; IPCC 2012; Mitchell and Harris 2012). The outcome of this transformation should be that new ecological, economic and social structures are created (Folke 2006). Other perspectives on resilience include socio-ecological and community resilience, where the former involves individual households or community members having the capability to adapt, to change, to anticipate or to respond—to shock and (future) uncertainty (Béné et al. 2014). In this vein, resilience is both about capability and agency, where people rather than systems and processes are at the center of the resilience effort. Proponents of community resilience suggest that communities are resilient if they survive a major disaster and return to a 'new normal state,' which usually involves positive change. Critical determinants of community resilience are the ability of the community to prepare, plan, absorb and recover (Cai et al. 2016; Cheer and Lew 2018; Ilies 2018; Stockholm Resilience Centre 2015). Community resilience is therefore closely entwined with the extent of vulnerability and adaptability of community households, so that 'enhancing community coping capacity and livelihoods can allow them to make appropriate choices within the context of their environments' (Manyena 2006, 434).

Thus, while resilience is touted as a universally understood concept, there is a high degree of ambiguity and difference in both meaning and application, so that according to Cutter et al. (2008 cited in Hall 2018, 19):

> Lingering concerns from the research community focus on disagreements as to the definition of resilience, whether resilience is an outcome or process, what type of resilience is being addressed (economic systems, infrastructure systems, ecological systems or community systems), and which policy realm (counterterrorism; climate change; emergency management; long-term disaster recovery; environmental restoration) should be the target.

Some have argued that it is its very ambiguity that has led resilience to be a mobilizing metaphor, bringing together a smorgasbord of interest groups and political agendas with the primary objective of 'strengthening resilience' (Pain and Levine 2012). Contrariwise, others have commented that the variety and ambiguity have rendered resilience useless and unable to allow for comparative analysis (Bahadur et al. 2010; Brand and Jax 2007). Three other critiques are worthy of consideration. One is that with a few exceptions, resilience definitions tend to focus on systems rather than actors (Davidson 2010). In other words, is resilience about the nature of the system or the nature of the people in the system (Manyena 2006)? A similar question has been articulated by the Department for International Development (DFID 2011) when they ask *resilience of what*? To approach resilience from a purely systems perspective would be tantamount to crafting a development agenda that considers only the policy aspect of development. One school of thought is that neither systems nor human actors should be privileged above the other for:

> While human beings should be at the centre of any resilience programme, they do not live in a vacuum, but instead are part of systems that impact on losses and the locality's ability to deal with them. (Manyena 2006, 444)

To render communities and individuals invisible or subordinate in resilience adaptation efforts, is to deny the strategies and capacities for coping, accommodating, absorbing and transforming that have been built up over centuries (Bec et al. 2018; Ilies 2018), which include but are not limited to indigenous intelligence, cultural heritage, occupational multiplicity, flexible specialization and lived experiences (Baldacchino and

Bertram 2009; Manyena 2006; Poon 1993). Exemplifying this, Ilies (2018) reports on how the indigenous mode of house construction served as a facilitator for resilient thinking as the very tradition spurred the development of ecotourism and cultural tourism in an area where employment opportunities were precarious and seasonal.

Second, is that ecological, community and to some extent, socio-ecological renderings of resilience have tended to overlook issues related to power and agency (Davidson 2010). Any contemporary review of the community development literature highlights that an understanding of power relationships is critical to robust analyses of social systems and institutions. Powerful actors can determine which community groups survive and recover from the shocks and which groups do not. Further, individuals and communities have a right to choose or self-determine, even poor people, the extent and ways in which they cope, adapt and accommodate external or internal shocks. In fact, resilience definitions that fail to acknowledge the role of power should also conclude that transformation which is potentially an outcome of resilience can be held back by power structures and actors (Baldacchino 2020; Coulthard 2012). Irrespective of whether resilience is seen as process or outcome, to omit the role of power in *what* and *who* is involved in resilience strategies, is to romanticize the concept and the way regional and international donor funding is often used to shape the resilience agenda at the local level. This consideration is pertinent for tourism development in SIDS which is often driven by foreign direct investment and international principals such as foreign airlines and tour operators (Britton 1982; Brohman 1996). Very often the resilience strategies that are adopted by businesses and political actors are aimed at ensuring that businesses and related networks remain intact after any disaster—with employees and small tourism businesses left 'twisting in the wind.' This statement by a representative of the international donor community in the wake of Hurricane Gilbert highlights how winners and losers emerge and how distributive resilience can occur.

> Although working groups were convened with responsibility for "Environment and Conservation" and for "Agriculture" (which presumably included fisheries), the Regional Coordinating Unit of the Caribbean Environment Programme, UNEP, felt that special attention should be paid to coastal and marine resources, because of their importance to the economy of this island state. (UNEP 1989)

More recently, the news item about the Government of Barbados compensating visitors to Barbados whose PCR tests were delayed, with the consequence of a longer quarantine, bears testament to how efforts at building destination brand resilience can favor others and disadvantage 'the other' (Clarke 2021). In contrast, many hotel employees have been unceremoniously dismissed or are awaiting compensation due to property closures (Smith 2020).

The final critique is that resilience is seen as a normative concept and therefore intrinsically good, leading always to a positive outcome (Baldacchino 2020; Hall 2018; Manyena 2006). Béné et al. (2014, 607) note that 'in relation to individuals or communities, resilience is not necessarily positively correlated with well-being; some households may have managed to strengthen their resilience but only at the detriment of their own well-being or self-esteem.' In other words, choices are sometimes made by individuals that would ensure their survival (resilience) but move them to the margins of poverty. In this regard, resilience can be construed as a trade-off between survival and well-being (Davidson 2010). Pain and Levine (2012) also point to the distributional aspect of resilience where project interventions to build resilience can result in some households and communities being winners and others losers. Theoretically then, resilience is inclusive but in practice it can have exclusionary outcomes. Failure to treat with the political, social justice, and trade-off elements of the resilience discourse can further entrench inequity, create false expectations and lead to undesirable outcomes (Robinson 2020).

This point is important for tourism resilience analysis especially where SIDS are concerned. With small populations, limited landmass and fragile environments, tourism development in SIDS is often faced with choices (implicit or explicit) concerning displacement of persons and industries, redirecting water from communities to green golf courses or removing mangroves that protect the shoreline in order to erect resorts on the beach front (Mowforth and Munt 2003). These political decisions geared toward employment stimulation and foreign exchange revenues have the inadvertent effect of creating an intrinsic level of economic and environmental vulnerability that has negative resilience consequences.

Notwithstanding these dilemmas, including the definitional haziness of the resilience concept, some common characteristics or determinants of a resilient system have been proposed and include notions of 'resistance, flexibility, diversity, inclusion and participation, recognition of social values, acceptance of uncertainty and change, learning and innovation, governance and a recognition of social processes such as community cohesion, good leadership and individual and collective action to cope with change' (Béné et al. 2014, 611–612). Other authors have commented that necessary prerequisites of resilient systems are information, communication and knowledge among social actors, strong institutional and community organizations, emergency preparedness and political stability (Cardona 2004 cited in Manyena 2006).

Dimensions of Resilience Responses

The attractiveness of resilience to policy makers and scientists (social and natural) as a framework around which individual and societal vulnerabilities can be identified and strategic interventions planned and implemented has led to the articulation of three dimensions of resilience, namely absorptive resilience, adaptive resilience and transformative resilience (Béné et al. 2012). Absorptive resilience pertains to a community's or society's ability to withstand the shocks of an event without any significant alteration to its state and functioning. This ability it is argued is largely because of sound planning, community (re) education and infrastructural strengthening through design which prevent or minimize disruption.

A relevant example of sound planning would be Jamaica where since the 1970s the island had established the National Disaster Committee (NDC), chaired by the then Prime Minister, Edward Seaga, comprising a multi-sectoral and ministerial team, including the Jamaica Fire Brigade (JFB), the Commissioner of Police and the Chief of Staff of the Jamaica Defense Force (JDF). The task of this Committee was 'to guide the Prime Minister on the planning and implementation of all measures considered necessary or desirable to plan for and counter the effects of disasters' (Office of Disaster Preparedness and Emergency Management 2019, n.p.). However, this institution became dormant during the 1990s, but was resuscitated in 1989 under the Chairmanship of Prime Minister Michael Manley. It is axiomatic that the reconstitution of this entity was a response to Hurricane Gilbert which resulted in '1.6 billion (U.S) or approximately two thirds of the GDP' (Davenport 1990, 53). Contrastingly, after Hurricane Andrew, which was a category 3 hurricane that hit The Bahamas in 1992, legislation was passed to improve the building code to withstand this level of hazard (Government of Bahamas 2003). The effect of this was that the housing stock was able to endure subsequent similar hurricanes. These two examples raise a number of questions around resilience governance, but more keenly demonstrate that absorptive resilience has to be constantly monitored, as the wider context of political cycles can underdo (or strengthen) previous progress made (Schweizer and Renn 2019).

When the absorptive capacity is exceeded, individuals and communities can also demonstrate adaptive capacity or resilience where they make adjustments, temporary or permanent, in order to continue functioning (Bene et al. 2012; Cutter et al. 2008). These adjustments can be incremental and over time can be quite insidious to the extent that some level of imperceptible trade-off is occurring. For instance, after the first three months of the COVID-19 pandemic, many households in the Caribbean began planting and growing their own food such as herb, vegetables and root crops (Doodnath 2020). This was a response to loss of employment in some instances, limited mobility and a means of coping psychologically with the 'cabin fever' associated with lockdown measures of many countries (The Story Editor 2021). As a policy response to encourage food security and ensure quality nutrition during the pandemic, governments have also sought to actively encourage home gardening initiatives by partially financing supplies such as small gardening equipment and seedlings (Lewis 2020). In this instance, the short-term trade-off was a fall in revenues by supermarkets for certain food items, but this may well continue as persons become accustomed to 'eating what they grow.'

Once the thresholds of adaptive resilience have been breached, transformation takes place in systems and societies in a way that the structures of society are irrevocably altered. Some have argued (cf Folke 2006) that transformative resilience challenges the status quo as the changes invariably lead to a permanent disruption of the values, beliefs, institutions and assumptions that were communally and formerly embraced. Transformational shifts are however, not without transactional costs and risks and 'the more you change the higher the transactional costs' (Bene et al. 2014, 602). For instance, the acceleration of the implementation of touchless technologies and physical distance measures in many tourism and hospitality businesses and organizations such as airports, hotels and restaurants in response to consumer preferences for virus minimizing engagements, has been reportedly quite costly. In spite of their financial losses, airlines reportedly doubled investments into the development of automated passenger self-boarding biometric and ID documentation systems hoping to reach implementation levels of 82% by 2023 up from

39% at present [Société Internationale de Télécommunications Aéronautiques (SITA) 2021]. Physical distancing measures have also challenged revenue gains by further restricting the already suppressed demand. IATA (2020) argues that blocking the sale of middle seats on airlines as a physical distancing measure would reduce available passenger capacity to approximately 62%, well below the average industry breakeven level of 77%, and warrant substantial increases in air fare between 43 and 54% to maintain breakeven levels. It is unlikely that businesses in tourism and hospitality will ever return to previous operating procedures even as they deliver service in an industry that is 'high touch,' with the cost of these transformations being evidently borne by the consumer. At the macro-level, questions have emerged around the probability of Caribbean or more broadly SIDS' governments, actively pursuing other economic activities or more radically, moving away from tourism in efforts to secure transformative resilience for more sustainable futures. The official promulgation and attendant actions by the government of St. Kitts and Nevis (SKN) in 2005 that 'sugar will no longer lead the economy of SKN but tourism' (Morton et al. 2010, 122) suggests that bold action for transformative resilience is indeed possible.

The usefulness of articulating these various dimensions of resilience should not be understated, but a similar caution is also necessary, and that is, this triad is not as linear as is conceptualized. In other words, it is unlikely that at the societal level that resilience responses move through these various stages in a step-by-step manner when there is an unexpected event. Rather, *within* societies all of these resilience dimensions might be simultaneously triggered as a consequence of an internal or external shock. In this regard, the level of vulnerability of individual and households *and* the level of preparedness and stability of the state would serve as the mediating variable. Hurricane Irma which devastated French St. Martin in 2017 brings into sharp relief the multidimensional effects of one hazard. In the wake of this disaster, some residents were looting businesses, while others had to find themselves in hurricane shelters temporarily and still others remained in their homes until the winds and rain abated (BBC News 2017). The estimated total insured damage was reported to be 1.176 billion euros (Jouannic et al. 2020). The French government led by President Macron responded quickly setting up an inter-ministerial team tasked with coordinating disaster relief and plotting a long-term reconstruction vision for the overseas territory in line with the President's vision which was: 'the island must be rebuilt as a "model" to withstand future weather patterns. I don't want to rebuild Saint Martin as it was ... We have seen there are many homes that were built too precariously, with fragile infrastructure. The geography of the homes was not adapted to the risks' (Al Jazeera 2017, n.p.).

The long-term vision for a sustainable rebuilding of St. Martin has been thwarted by a number of challenges *inter alia* structural dysfunctions in government, local capacity, technical skills, lack of local knowledge of foreign experts, cumbersome financing mechanisms and lack of public consultation (Jouannic et al. 2020). While the French government can certainly not be accused of a narrow or linear resilience focus (absorptive versus adaptive vs transformative) that often bedevils many countries and which results in planning and scale limitations (Holladay 2018), the case highlights the absence of critical requirements needed to move from absorption to transformative resilience.

Island Vulnerability and Resilience

As discussed in detail in Chapter 2, the traditional narrative on small island states is that they have a number of sustainable development challenges that include remoteness, limited resources, susceptibility to natural disasters, over reliance on a few exports and fragile environments (Kronenberg and Khor 2016). The United Nations Conference on Environment and Economic Development (1992) avers that these endemic island characteristics have led to high communication, energy, input and transportation costs and diseconomies of scale. These tropes around small islands have inevitably resulted in the formation of an easy relationship between the islands and vulnerability literature, where SIDS are emphatically, economically and environmentally vulnerable and require concomitant resilience capacities.

Resilience has therefore become both synonymous with and antithetical to vulnerability and the solution to island vulnerability and climate adaptation (Brinklow 2020), with climate change finance feeding the resilience discourse (Baldacchino 2020). A quick review of the international donor-funded project discourse would undoubtedly highlight some key words—island vulnerability, climate change, adaptation, mitigation and resilience. Baldacchino (2020, n.p.) argues that 'islands need to wrest themselves out of the ontological strictures of 'the resilience is the answer to vulnerability' narrative and claim a new discourse for navigating economic and environmental shocks.'

The development of a vulnerability index for SIDS has reinforced the shackling of small island states to the vulnerability narrative (Briguglio 2016, 2020). The thesis is that SIDS are exposed to varying levels of risk (vulnerability) because of the inherent characteristics identified above. The exception to this has been Singapore—referred to as the 'Singapore paradox' where despite the country's vulnerability it still has a high GDP per capita, mainly as a result of strong policy governance (Briguglio et al. 2009). Other paradoxes include 'Malta's March,' 'Mauritius Miracle,' and 'Aruba's Ascent' (Peterson et al. 2017). In contrast, the majority of Caribbean, Indian Ocean and Pacific Islands are economically vulnerable as a consequence of overdependence on tourism, Citizenship by Investment Programme (as in the case of Vanuatu) and remittances, to name a few.

Table 3.1 provides a snapshot of events (natural, health, epidemics, economic) that have exposed the vulnerabilities of the Caribbean region from 1995 to 2021, and the significant human, environmental and economic losses that have been incurred. Current estimates of declines in travel in some of the Caribbean's main tourism source markets as a result of the current COVID-19 pandemic show: Euro Area (−7.5%); UK (−6.5%); Canada (−6.2%); and USA (−5.9%) (International Monetary Fund 2020). And, the UNWTO has predicted that international tourist arrivals will decline by 20–30% which translates into US$300–450 billion in receipts (World Tourism Organization 2020). It is axiomatic that many tourism-dependent SIDS will be hit hard by the effects of this virus, due to the high level of unemployment in many of the source markets. Delivering a grim report on the economic performance of Barbados for the January–March 2021 quarter, the Governor of the Central Bank of Barbados stated 'the short-term cost of the pandemic in terms of jobs and incomes has been considerable. The weakened tourism sector, together with the fall out in other economic activities, and depressed private spending contributed to an estimated 19.8% fall in output. Tourism output plunged by an estimated 96%' (Central Bank of Barbados 2021, 1–2).

The other perspective on vulnerability does not view economic vulnerability as a disadvantage *per say* but as a natural phenomenon with which SIDS are faced and which can yield both positive and negative consequences (Baldacchino 2015a, 2015b). Island responses to economic shocks include agile and entrepreneurial responses, occupational multiplicity, focusing on economies of scope rather than scale,

Table 3.1 Shock events in the Caribbean 1995–2021

Year	*Event*	*Island*	*Damages*[1] *(US$)*	*Total affected*
1995	Volcanic eruption	Montserrat	–	5,000
2003	Typhoid fever	Haiti	–	200
2004	Hurricane Ivan	Cayman Islands	3.4B	N/A
		Grenada	889M	60, 00
		Jamaica	595M	350,00
	Storm Frances	Bahamas	1B	8,000
2005	Hurricane Dennis	Cuba	1.4B	2,500,000
2008/2009	Global financial crisis	Barbados	280M lost tourism revenue (Worrell 2010)	N/A
2008	Hurricane Gustav	Cuba	2B	450,000
2010	Earthquake	Haiti	8.5B (Trocaire 2012)	N/A
	Cholera outbreak	Haiti	–	513,997
2012	Hurricane Sandy	Jamaica	16M	215,850
		Haiti	254M	201,850
2014	Chikungunya virus disease	Haiti	–	39,343
2015	Cholera outbreak	Haiti	–	20,000
	Sargassum outbreak	Caribbean Region	120M clean-up costs (Christie 2020)	–
	Hurricane Erika	Dominica	482M	228,594
2016	Zika outbreak	Jamaica	112M in income foregone (World Bank 2016)	N/A
	Hurricane Matthew	Cuba	2. 6B	190,000
		Haiti	2B	N/A
	Drought	Haiti	84M	N/A
2017	Hurricane Maria	Puerto Rico	US. 68B	750,000
		Dominica	1.5B	71,393

(continued)

[1] Represents damages unless otherwise stated.

Table 3.1 (continued)

Year	*Event*	*Island*	*Damages(US$)*	*Total affected*
	Hurricane Irma	St. Martin/ St. Maarten	4.1B (St. Martin); 2.5B (St. Maarten)	11,400 (St. Maarten)
		BVI	3B	N/A
		Antigua and Barbuda	250M	95% of homes in Barbuda destroyed (OCHA-UN 2018)
2019	Tropical Storm Dorian	Bahamas	3.4B	14,940
2020–Current	COVID-19	Bahamas	3Billion—Estimated loss in tourism revenue in 2020 (Hartnell 2021)	N/A
2021	Volcanic eruption	St. Vincent	TBC	TBC

Sources Damages and total affected data is from CRED (2020) unless otherwise stated (Christie 2020; Hartnell 2021; OCHA-UN 2018; Trocaire 2012; Wallemacq and House 2018; World Bank 2016; Worrell 2010)

and flexible specialization (Baldaccino and Bertram 2009). The impasse between the Briguglio-Baldacchino schools of thought on vulnerability has led to the development of two vulnerability-resilience strands of thought, where Briguglio (2016) emphasizes good political, economic, social and environmental governance as prerequisites for resilience and Baldacchino and Bertram (2009) focus attention on social resilience or the innate resourcefulness of islanders to overcome exogenous events. In my opinion, the perceived irreconcilability of these two positions seems forced; certainly, one can argue that both resilience responses should be simultaneously privileged and embraced. Thus, in discussing tourism's resilience in vulnerable territories, one cannot minimize the fact that while SIDS are indeed vulnerable to external shocks and natural and man-made disasters, some have been able to counter these vulnerabilities through communal resources, traditional beliefs and a variety of resilience strategies (Scheyvens and Momsen 2008). Similarly, in the aftermath of Hurricane Irma, the British Virgin Islands (BVI) was able to recover quickly largely because the institutional policy mechanisms (CDEM[2]

[2] CDEM—Comprehensive Disaster Emergency Management.

Policy, CDEM legislation, work programme, reporting, monitoring and evaluation system) were in place and well supported.

Examples of Caribbean SIDS' ability to adapt, resist and recover from the effects of shock or change are perhaps best known in the area of natural disasters. For example, in the wake of Hurricane Ivan in Grenada and Hurricane Maria in Dominica, Caribbean families voluntarily hosted primary and secondary school students who had imminent qualifying examinations. Less well known is the region's institutional resilience with respect to the Caribbean Community (CARICOM), an institution that has had several starts and stops due to nationalistic and sovereignty claims of its member states, but which has survived for the last 47 years (Grenade 2008); even giving birth to other key bodies such as CDEMA, CCCCC, CARPHA and the CCJ- institutions[3] that function and collaborate to strengthen the region's absorptive, adaptive and transformative capacities.

The Resilience Framework in SIDS

Where the development and natural sciences literature on resilience addresses escalating dimensions of resilience, research on island resilience employs a more PESTE-type framework. So that, economic resilience refers to 'the policy-induced ability of an economy to recover from or adjust to the negative impacts of adverse exogenous shocks and to benefit from positive shocks' (Briguglio et al. 2008, 5). Figure 3.1 proposes a vulnerability-resilience nexus where resilience is seen as the ability to cope, withstand and bounce back from unexpected shocks and is only possible through policy and institutional buffers such as sound structural policies, compliance, investment in infrastructure and social services, strong institutions and good governance (Briguglio 2016; Cabezon et al. 2016). Based upon their policy architecture, SIDS can determine where they fall in the matrix. Interestingly, island entrepreneurship and other strategies cited earlier are omitted from the economic resilience solutions, even though studies have shown that in the face of economic shocks, the entrepreneurial spirit of islanders has comes to the fore, helping them

[3] CDEMA—Caribbean Disaster Emergency Management Agenc.
CCCCC—Caribbean Community Climate Change Centre.
CARPHA—Caribbean Public Health Agency.
CCJ—Caribbean Court of Justice.

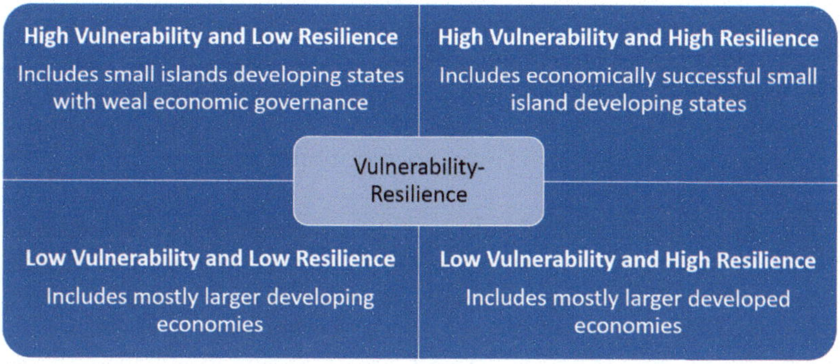

Fig. 3.1 Vulnerability-resilience nexus (*Source* Briguglio 2016)

to weather unemployment and associated effects (Butler 2018; Peterson et al. 2017).

Disaster resilience is another dimension of resilience and is defined as 'the ability of countries, communities and households to manage change by maintaining or transforming living standards in the face of shocks or stresses -such as earthquakes, drought or violent conflict-without compromising their long-term prospects' (Department for International Development 2011, 6). The emphasis on 'maintaining and transforming' social systems is somewhat akin to the absorption and transformative responses proposed by Bene et al. (2014) albeit, without acknowledging that some type of trade-off might be necessary to achieve transformation. DFID further proposes a resilience framework as a way to understand the level of resilience that exists within a particular country, community or household. The four elements include:

1. *The context*—what type of political, social, economic, technological structures exist.
2. *Disturbance*—what are the types and intensity of the shocks and stresses that impact the vulnerability of the context. These can include natural hazards, conflict, insecurity, pandemics.
3. *Capacity to deal with disturbance*—this is related to the country's or community's level of exposure, level of sensitivity and adaptive capacities. For example, if a particular locale has never endured a hurricane (exposure), many persons might not know what actions

to take (sensitivity) and the organizations and institutions might be ill prepared to respond and cope with the effects (adaptive capacity).

4. *Reaction to disturbance*—the level of vulnerability identified in stages 1–3 will determine whether the outcome will be that the society survive, cope, recover, learn or transform.

This framework is undoubtedly useful for risk mapping and responses in many contexts, including SIDS; however, the volatility and rapidly changing hazard landscape suggest that unmitigated adherence to its tenets would be at best unwise. In the contemporary environment, multiple types and scales of hazards are occurring simultaneously, as was the case in Sendai, Japan, when a 7.2 magnitude earthquake was followed by multiple tsunami waves. Similarly, in the grips of managing the COVID-19 pandemic, St. Vincent and the Grenadines (SVG) was subject to the eruption of the La Soufriere volcano, and a few weeks later, to heavy rainfall resulting in flooding, landslides, destruction of home and infrastructure (Mendes-Franco 2021). The overwhelming nature of the multiple hazard environment especially in SIDS, needs to address deeper structural, cultural and environmental issues than is covered in this framework. In this vein, Wilkinson and Peters (2015, 50) cogently state that 'the challenges of underpinning projections of future climate change and its augmentation of extreme events, along with the development context in which these events will play out, requires new approaches to supporting resilience.'

Finally, the notion of social and cultural resilience alluded to above infers that island states and islanders possess an inherent 'flexibility, resourcefulness and strategic response to threats and opportunities that allow them to surmount economic and other exogenous shocks that they experience (Baldacchino and Bertram 2009, 144). Drawing upon her own experience, Brinklow (2020) takes the view that 'islandness' creates its own resilience machinery as it is not so much a matter of coping, but a matter of survival. The possibility of survival is engendered through strong communal and kinship ties, a shared mythology and the creation of other forms of social capital, and as advanced by Cheer and Lew (2018) is a precursor for building community resilience. This concept of 'social glue or cultural capital' as termed by Baldacchino and Bertram (2009) has found expression in the Caribbean region in times of adversity. Within days of the eruption of the La Soufriere volcano, the

Caribbean Community and its diaspora mobilized to assist SVG with evacuation, food, medical supplies and so on. These gestures have resonated throughout the region in moments such as these and underscore the instinctive rapid response capabilities of the region. Figure 3.2 is a letter sent to Prime Minister Mottley of Barbados to Prime Minister Rowley of Trinidad and Tobago thanking him for his country's generosity in the face of the unforeseen-unprecedented ash fall in Barbados as a result of the eruption. These actions are in some ways paradoxical as individually each of these independent Caribbean islands fight to determine sovereign economic resilience strategies (as opposed to cooperative) but in times of natural crises, social resilience rises to the top.

The foregoing articulation of the various dimensions of resilience and their association with narratives of survival, resourcefulness and coping relative to SIDS align more closely with absorptive and adaptive resilience capacities, and I argue, are somewhat truncated in that they do not go far enough to address transformations such as improvements in quality of life, social services, institutional strengthening, funding requirements for vulnerability and capacity building in SIDS (CDEMA and CARICOM 2018). The architects of the Caribbean Pathway for Disaster Resilience in the Caribbean Community state that for the region absorption and adaptation strategies are not sufficient to '*bounce forward quickly*' in a manner that reduces susceptibility (increased liability to additional harm) to the impact of the hazard (CDEMA and CARICOM 2018, 3). According to the Pathway document:

> true resilience for the Caribbean will require transformation! Transformation is not business as usual but essentially involves challenging the existing status quo including the systems and power relationships which cause the accumulation of vulnerability and which must be identified and changed, if root causes are to be addressed and societies are to be transformed. (p. 3)

In accordance with this transformative resilience approach, the Pathway identifies 5 Pillars underpinned by a set of key foundational and reinforcing elements which facilitate their delivery:

I. Social Protection for the Marginal and Most Vulnerable
II. Safeguarding Infrastructure
III. Enhancing Economic Opportunity
IV. Environmental Protection

PRIME MINISTER
BARBADOS

April 19, 2021

Dear Keith,

I write to you at this time when Barbados is experiencing its most recent crisis, brought about by the eruption of the La Soufriere Volcano in St Vincent and the Grenadines; and which has caused our country to be covered in ash thereby leading to some measure of disruption, in addition to potential health and environmental hazards. This letter serves to convey heartfelt appreciation to the Government and people of the Republic of Trinidad and Tobago for your act of kindness and cooperation when you readily responded to our plight at a moment's notice, arranging for brooms, goggles and protective overalls to be sourced throughout Trinidad.

Events such as this underscore how inextricable linked we are as CARICOM Member States. For it was our proximity to St. Vincent which sent us the ash and, by the same token, our proximity to Trinidad and Tobago which enabled us to receive these much needed supplies, transported by both your Air Guard plane and the Regional Security System's Air Wing.

It is indeed heartwarming to me that although you are currently in isolation, following your recent COVID-19 positive test, you were thoughtful enough to reach out and offer this assistance. I wish you a speedy recovery on behalf of my Government and all Barbadians, and reaffirm my commitment to the tenets of regional cooperation. Prime Minister thank you and God bless.

Yours sincerely

Mia Amor Mottley, Q.C., M.P.

The Honorable Keith Rowley
Prime Minister
Republic of Trinidad and Tobago

Prime Minister's Office, Government Headquarters, Bay Street, St. Michael, Barbados
Tel: (246) 535-5350/5348 • Fax: (246) 535-5656 • E-mail: pmbarbados@barbados.gov.bb

Fig. 3.2 Thank you letter to Prime Minister Rowley from Prime Minister Mottley

V. Operational Readiness and Recovery

Of note is the recognition by the authors of the role of power in the region's progress to Caribbean resilience, so that powerful actors such governments and private sector titans can make, break or stall this vision, once the required transformations are not in alignment with the 'needs' of the political cycle or the current business model. Further, the right plans and structures might be woefully inadequate to effect regional transformation, if time and money are not spent on changing human behaviors. Reflecting a similar sentiment, Robinson (2020) posits that while there is an understanding of resilience there seems to be a disconnect between knowledge, attitudes and practice. Overcoming this barrier would require both a stronger relationship between science and policy and a better understanding of the institutional and cultural factors that impede effective resilience strategies.

The lingering question then becomes 'how does transformative resilience align with a region that does not appear to see a future beyond tourism?' Bene et al. (2014) have suggested that transformative resilience has the highest transaction costs and perhaps, it is for that reason island governments seem unable and unwilling to address the possibility of a new economic model. Thus, pandemic resilience efforts are geared toward upskilling, reskilling and new skilling tourism sector employees, establishing vaccination travel protocols for visitors, offering digital nomads an opportunity to live and work in some islands through various initiatives, providing economic stimulus packages for operators across the sectors and so on—all *in preparation for* when travel resumes. The Reimaging Tourism conference hosted jointly by the Inter-American Development Bank (IDB) and the Barbados Ministry of Tourism in March 2021, axiomatically reinforced the resetting of the sector 'as the central pillar of the national economy' with the island nation receiving US$20 million in ID donor funds to increase the resilience of the tourism industry (Archibald 2021).

Thompson (2020, n.p.) is not so optimistic in his outlook for tourism, suggesting that 'tourism is a dead horse. More carrots or bigger sticks will not work. We need to think beyond tourism.' His proposed solution is for Barbados to leverage its climate in more productive ways, particularly through the development of an International Resident sector where high earning individuals would relocate to work in Barbados. Such an initiative was announced on 30 June 2020 under the rubric of the Barbados

Welcome Stamp Initiative; a similar initiative, Bahamas Extended Access Travel Stay programme was also launched in the Bahamas. The problem with these strategies is that they are still premised upon an overdependence of tourism, albeit an extended stay version, once 365 days have not expired. An International Monetary Fund (IMF) opinion piece recommends a combination of strategies for the post-pandemic Caribbean including mobilizing concessional financing to build resilience against future shocks, shifting to more niche tourism products that are low-density and high value-added and that create local intersectoral linkages. Diversification strategies to enhance resilience should focus on the 'blue economy,' shipping, fisheries and aquaculture. Underpinning all these economic activities should be technological innovation to improve efficiency and reduce costs (Srinivasan et al. 2021). What emerges from much of these COVID-19 discussions and recommendations are that they all align with an economic resilience ethos that continues to unquestioningly reinforce the overdependence on tourism by many Caribbean island states. The expected outcome in this regard is likely to be a continuation of unsustainable vulnerability.

Tourism Resilience

The foregoing has sought to shine light on the epistemological evolution of resilience and the concomitant frameworks, contestations and determinants that have been proposed to stimulate thought and lead to action. However, within the tourism resilience literature very little of this debate is unpacked or distilled; instead, an a priori engineering perspective is adopted where the focus is on rebuilding the destination to welcome back visitors (Hall 2018; Ibanescu et al. 2020). Framed in this way, Destination Management Organizations (DMOs) have a crucial role to play in ensuring that the destination bounces back after a natural disaster or a man-made crisis event (Basurto-Cedeno and Penington-Gray 2016). Adaptation strategies that supported engineering and by extension economic resilience include tax incentives, grants, retention of jobs, public sector subsidized training programmes, destination promotional campaigns and staycation programmes to name a few (Holladay 2018; Ritchie 2017). Depending on the nature and the scale of the destruction, adaptation strategies for resilience also feature green initiatives, new or improved coastal infrastructure and the adoption of green technology and alternative energy (Kruse et al. 2013).

What has therefore emerged are three research strands underpinned by economic and engineering resilience which are destination resilience, business resilience and community resilience. Destination resilience as alluded to above is the effort by tourism-dependent governments and vested private sector and civil society to ensure that 'the destination' rebounds from negative shocks as quickly as possible aided by a number of government concessions. Impeding the adaptation efforts are funding inadequacies, ignorance of destination ecosystems and capacity, poor stakeholder relations and the failure to include tourism in broader national goals (Bangwayo-Skeete and Skeete 2020; Basurto-Cedeno and Penington-Gray 2016; Ibanescu et al. 2020). Business resilience is intricately tied to destination resilience given the critical role that tourism businesses—first and second tier—play in providing visitor services. In times of crises or shocks, tourism businesses especially in tourism-dependent jurisdictions reflect varying levels of business resilience with respect to their capacity to survive, adapt and innovate (Dahles 2018). One school of thought has argued that small tourism businesses show a higher degree of resilience because of the existence of functional redundancies such as 'diversity, self-organization, self-correction, and local control' (Alebaki and Ioannides 2018, 158). For businesses and destinations, Kopp et al. (2014) propose three types of adaptation needed in securing business continuity, namely technical adaptation that incorporates new and traditional technologies; business management adaptation that is aligned with marketing communications and pricing strategies; and behavioral adaptation associated with changing tourist behavior through information.

Finally, community resilience has emerged arguably in tandem with sustainable tourism ideas around local empowerment, participatory development and spreading the benefits of tourism. Community resilience as an analytical framework intersects with the economic and socio-ecological perspectives on resilience in that while it focuses on the 'social survival processes that are put into action by local communities to address the negative social and economic impacts experienced during crisis' (Imperiale and Vanclay 2016 cited in Cheer and Lew 2018, 69), there is also a sense in which tourism is used as an economic resilience tool to bolster dying communities (Butler 2018; Ibanescu et al. 2020; Ilies 2018). Recognizing that community resilience is not an apolitical concept, Cheer and Lew (2018) have called for more interventionist policies beyond what government and policy offer, in seeking to build community resilience.

Therefore, with the exception of Sheller (2020) very little consideration is given to moving the boundaries of resilience away from tourism to consider other economic options in the variety of study contexts proposed. A similar argument is posited in the foreword of this book where Baldacchino calls on Caribbean SIDS to diversify within and beyond tourism. Thus, the treatment of resilience is very reminiscent of the 'sustainable tourism development versus sustainable tourism' discourse where, as argued by Butler (1999), the emphasis of the latter is on sustaining tourism. This approach to tourism resilience is surprising for two reasons. Firstly, because of the implicit power dimension of tourism, especially in developing countries and in SIDS, where these destinations made up of many small tourism businesses, are often price takers and not price makers. Any unquestioning approach to the praxis of tourism resilience is 'a simple return to normality where dominant interests regain control and leave wider socio-cultural concerns unaddressed' (Hall 2018, 25) and perpetuate the cycle of winners and losers (Sheller 2020).

Secondly, is that tourism is often regarded as a system comprising very interrelated components—the tourists, geographical elements and the industry (Leiper 1979). The tourism system is therefore by no means a closed system, meaning that a shock in one part of the system can have deleterious effects on the other part(s), making the entire industry very vulnerable. An understanding of the interrelated and complex nature of the tourism system should easily lend to research on tourism resilience that centers and builds upon the socio-ecological resilience approach that accepts non-linearity, dynamism, uncertainty, adaptability and transformation, as well as the role of power among a myriad number of actors, when dealing with shocks.

The Adaptive Cycle Model

Notwithstanding the foregoing, there seems to be a tacit acceptance of the socio-ecological approach to resilience by tourism researchers evidenced by the oft-cited adaptive cycle model which rejects the equilibrist notion of a balanced and stable system (Holling 2001). A conflated version of this model also employs Butler's (1980) Tourism Area Life Cycle. Applied to tourism, the adaptive cycle proposes that tourism systems respond to changes by moving through repeated cycles and loops (phases), where the speed of recovery is determined by the tourism system's ability or capacity to manage stress. In each of these phases, resilience is characterized by different economic accommodations, so that in the first phase, exploitation of resources leads to growth; the conservation phase is marked by stability, increasing rigidity and dependency on the resources, in the release phase there is chaos around the resource, making it vulnerable to shocks and in the final phase reorganization, where there is innovation and transformation or complete extinction (Holling 2001; Holladay 2018). Dahles (2018) points out that common to the socio-ecological approaches on resilience are functional redundancy and cyclical change so that 'as systems mature and resilience diminishes, their demise makes way for alternatives to develop and innovations to materialize' (p. 151). With reference to a remote tribal village in Taiwan, where tourism was used as a tool for building resilience, Wu and Wall (2018) concluded that the reorganization stage of the adaptive cycle has the highest potential for resilience because it is here that the system is most open to innovation. However, they caution that building adaptive resilience requires long-term commitment and trust especially in remote locales where the project donors often leave at the end, even though local capacity might not be adequately improved.

Worth mentioning is a study conducted by Bangwayo-Skeete and Skeete (2020) that quantitatively applies the adaptive cycle model in a comparative analysis of Barbados and Grenada. Using four decades of tourist arrival data, and citing various shocks experienced by both islands including nine global economic recessions, the September 11 terrorist attack and in the case of Grenada, Hurricane Ivan in 2004, the study found that both tourism economies went through the exploitation and decline phases but exhibited different resistance to shocks, with Grenada showing greater resilience than Barbados. The reasons for Grenada's quick rebound from these disruptions in terms of tourism growth had to do with the island's 'adequate international, regional and government support, macroeconomic restructuring, and political and social cohesion' (p. 19). Conversely, in Barbados, there was an absence of governance cohesion and integrated intersectoral communication. The application of this model to track resilience levels in SIDS is indeed laudable especially as it points the way to the possibility of transformative and regenerative strategies for these islands within and outside of tourism.

Summary Thoughts

This chapter has sought to highlight the contestations around the concept of resilience and the implications for tourism resilience in SIDS. What has emerged from the discussion is that tourism researchers have taken a case study application rather than a theoretical approach, so that thorny issues related to resilience 'of what' and resilience 'to what' have not been explicitly answered. These are important questions given the relationships of power that exist within SIDS and between SIDS and the international tourism system. In this regard 'bouncing back' might only mean cementing pre-existing inequalities, continued environmental degradation and maintaining poor sectoral linkages. Thus, adaptation and absorption while useful in some circumstances are likely to be the accepted resilience posture in many SIDS, where the focus will continue to be on destination, business and community resilience, as a response to shocks. Indeed, while transformative resilience or 'bouncing forward quickly' is the preferred option for a more sustainable Caribbean future, it is understandably not the easiest option as both costs and benefits would need to be borne by all and would require political, private sector and civil society buy-in to transform our livelihood options into a more diversified and sustainable future.

References

Al Jazeera. 2017, September 13. *Macron: France Will Rebuild St Martin Island after Irma*. Al Jazeera. https://www.aljazeera.com/news/2017/9/13/macron-france-will-rebuild-st-martin-island-after-irma.

Alebaki, Maria and Ioannides, Dimitri. 2018. Threats and Obstacles to Resilience: Insights from Greece's Wine Tourism. In *Tourism, Resilience and Sustainability: Adapting to Social, Political and Economic Change*, ed. Joseph M. Cheer and Alan A. Lew, 132–148. Abingdon, Oxford: Routledge.

Archibald, Yajaira. 2021, March 27. What's Next for Barbados' Tourism Industry? *Caribbean Development Trends. Inter-American Development Bank*. https://blogs.iadb.org/caribbean-dev-trends/en/whats-next-for-barbados-tourism-industry/.

Bahadur, Aditya V., Maggie Ibrahim, and Thomas Tanner. The Resilience Renaissance? Unpacking of Resilience for Tackling Climate Change and Disasters. CSR Discussion Paper No. 1, Strengthening Climate Resilience Programme, Brighton: Institute of Development Studies. https://www.fsnnetwork.org/sites/default/files/ids_resilience-renaissance.pdf.

Baldacchino, Godfrey. 2015a, July 28. Book Review: Building the Resilience of Small States: A Revised Framework. *The Round Table* 104 (4). https://doi.org/10.1080/00358533.2015.1064592.

Baldacchino, Godfrey. 2015b. Editorial: Small Island States and Territories: Vulnerable, Resilient, But Also Doggedly Perseverant and Cleverly Opportunistic. *Entrepreneurship in Small Island States and Territories*, 23–50. https://doi.org/10.4324/9781315764528-7.

Baldacchino, Godfrey. 2020. *The Challenges of Studying Resilience: A (Small) Island Perspective*. Paper Presented at Researching Resilience in Islands, Online, November 16–17, 2020. https://storymaps.arcgis.com/stories/c39f201ecb22417c9d2256adb045c352.

Baldacchino, Godfrey, and Geoffrey Bertram. 2009. The Beak of the Finch: Insights into the Economic Development of Small Economies. *The Round Table* 98 (401): 141–160. https://doi.org/10.1080/00358530902757867.

Bangwayo-Skeete, Prosper F., and Ryan W. Skeete. 2020. Modelling Tourism Resilience in Small Island States: A Tale of Two Countries. *Tourism Geographies*, 1–22. https://doi.org/10.1080/14616688.2020.1750684.

Basurto-Cedeño, Estefania Mercedes, and Lori Pennington-Gray. 2016. Tourism Disaster Resilience Scorecard for Destinations (TDRSD): The Case of Manta, Ecuador. *International Journal of Tourism Cities* 2 (2): 149–163. https://doi.org/10.1108/ijtc-01-2016-0002.

BBC. 2017. Hurricane Irma: Troops Deployed Against St Martin Looting. *BBC News*, September 8, 2017. https://www.bbc.com/news/world-latin-america-41203286.

Bec, Alexandra, Moyle, Brent, McLennan, Char-Lee and Keast, Robyn. 2018. Strategies for Building Community Resilience to Long-Term Structural Change in the Mackay and Whitsunday Regions of Queensland, Australia. In *Tourism, Resilience and Sustainability: Adapting to Social, Political and Economic Change*, ed. Joseph M. Cheer and Alan A. Lew, 167–183. Abingdon, Oxford: Routledge.

Béné, Christophe, Rachel Godfrey Wood, Andrew Newsham, and Mark Davies. 2012. Resilience: New Utopia or New Tyranny? Reflection About the Potentials and Limits of the Concept of Resilience in Relation to Vulnerability Reduction Programmes. IDS Working Papers 405: 1–61. https://doi.org/10.1111/j.2040-0209.2012.00405.x.

Béné, Christophe, Andrew Newsham, Mark Davies, Martina Ulrichs, and Rachel Godfrey-Wood. 2014. Review Article: Resilience, Poverty and Development. *Journal of International Development* 26 (5): 598–623. https://doi.org/10.1002/jid.2992.

Brand, Fridolin Simon, and Kurt Jax. 2007. Focusing the Meaning(s) of Resilience: Resilience as a Descriptive Concept and a Boundary Object. *Ecology and Society* 12 (1): 23. https://doi.org/10.5751/es-02029-120123.

Briguglio, Lino. 2016. Building Resilience in the Pacific Island Countries. In *Resilience and Growth in the Small States of the Pacific*, ed. Hoe Ee Khor, Roger P. Kronenberg, and Patrizia Tumbarello, 39–44. Washington, DC: International Monetary Fund.

Briguglio, Lino. 2020. *Two Approaches in Explaining Economic Resilience in Small Island States.* Paper Presented at Researching Resilience in Islands, Online, November 16–17, 2020. https://storymaps.arcgis.com/stories/c39f201ecb22417c9d2256adb045c352.

Briguglio, Lino, Gordon Cordina, Nadia Farrugia and Stephanie Vellas. 2008. "Economic Vulnerability and Resilience: Concepts and Measurements." WIDER Working Paper 55/2008, UNU Wider. https://www.wider.unu.edu/publication/economic-vulnerability-and-resilience.

Briguglio, Lino, Gordon Cordina, Nadia Farrugia, and Stephanie Vella. 2009, September. Economic Vulnerability and Resilience: Concepts and Measurements. *Oxford Development Studies* 37: 229–247. https://doi.org/10.14217/9781848598881-6-en.

Brinklow, Laurie. 2020. *From Cultural Cringe to Cultural Confidence: The Role of Islandness in Creating Resilience.* Paper Presented at Researching Resilience in Islands, Online, November 16–17, 2020. https://storymaps.arcgis.com/stories/c39f201ecb22417c9d2256adb045c352.

Britton, Stephen G. 1982. The Political Economy of Tourism in the Third World. *Annals of Tourism Research* 9 (3): 331–358. https://doi.org/10.1016/0160-7383(82)90018-4.

Brohman, John. 1996. New Directions in Tourism for Third World Development. *Annals of Tourism Research* 23 (1): 48–70. https://doi.org/10.1016/0160-7383(95)00043-7.

Butler, Richard W. 1980. The Concept of a Tourist Area Cycle of Evolution: Implications for Management of Resources. *The Canadian Geographer/le Géographe Canadien* 24 (1): 5–12. https://doi.org/10.1111/j.1541-0064.1980.tb00970.x.

Butler, Richard W. 1999. Sustainable Tourism: A State-of-the-Art Review. *Tourism Geographies* 1 (1): 7–25. https://doi.org/10.1080/14616689908721291.

Butler, Richard. 2018. Resilience in the Face of Changing Circumstances: Fair Isle, Shetland. In *Tourism, Resilience and Sustainability: Adapting to Social, Political and Economic Change*, ed. Joseph M. Cheer and Alan A. Lew, 116–131. Abingdon, Oxford: Routledge.

Cabezon, Ezequiel, Patrizia Tumbarello, and Wu Yiqun. 2016. Raising Potential Growth and Enhancing Resilience to Shocks. In *Resilience and Growth in the Small States of the Pacific*, ed. Hoe Ee Khor, Roger P. Kronenberg, and Patrizia Tumbarello, 15–38. Washington, DC: International Monetary Fund.

Cai, Heng, Nina Lam, Lei Zou, Yi Qiang, and Kenan Li. 2016. Assessing Community Resilience to Coastal Hazards in the Lower Mississippi River Basin. *Water* 8 (2): 46. https://doi.org/10.3390/w8020046.

Caribbean Disaster Emergency Management Agency (CDMEA). 2009. *Disaster Risk Management Strategy and Plan of Action for the Tourism Sector in the Caribbean Part II: Plan of Action*. Lodge Hill, Saint Michael, Barbados.

CDEMA and CARICOM. 2018, July. *Building a Caribbean Pathway for Disaster Resilience in the Caribbean Community (CARICOM)*. Barbados: Bridgetown. https://www.cdema.org/cdm11/Building_A_Caribbean_Pathway_For_Disaster_Resilience_In_The_Caribbean_Community_(CARICOM)_FINAL.pdf.

Central Bank of Barbados. 2021, April 29. *Review of Barbados' Economic Performance—January to March 2021*. Press Release. Central Bank of Barbados. http://www.centralbank.org.bb/Portals/0/Files/Review%20of%20the%20Barbados%20Economy%20-%20Jan-March%202021.pdf.

Cheer, Joseph M., and Alan A. Lew. 2018. Understanding Tourism Resilience: Adapting to Social, Political, and Economic Change. In *Tourism, Resilience and Sustainability: Adapting to Social, Political and Economic Change*, ed. Joseph M. Cheer and Alan A. Lew, 3–17. Abingdon, Oxford: Routledge.

Christie, Alison Binger. 2020, June 17. *Turning Trash into Treasure: Tackling Sargassum in Jamaica*. Inter-American Development Bank. https://www.iadb.org/en/improvinglives/turning-trash-treasure-tackling-sargassum-jamaica.

Clarke, Sherrylyn. 2021, January 10. Government Footing Hotel Costs for People Awaiting COVID Tests. *NationNews Barbados*. https://www.nationnews.com/2021/01/10/government-footing-hotel-costs-people-awaiting-covid-tests/.

Coulthard, Sarah. "Can We Be Both Resilient and Well, and What Choices Do People Have? Incorporating Agency into the Resilience Debate from a Fisheries Perspective." *Ecology and Society* 17, no. 1 (March 2012): 4. https://doi.org/10.5751/ES-04483-170104.

CRED. 2020. EM-DAT Public. *EM-DAT—The International Disasters Database*. Centre for Research on the Epidemiology of Disasters —CRED. https://public.emdat.be/.

Cutter, Susan L., Lindsey Barnes, Melissa Berry, Christopher Burton, Elijah Evans, Eric Tate, and Jennifer Webb. 2008. A Place-Based Model for Understanding Community Resilience to Natural Disasters. *Global Environmental Change* 18 (4): 598–606. https://doi.org/10.1016/j.gloenvcha.2008.07.013.

Dahles, Heidi. 2018. The Sustainability of Small Business Resilience: The Local Tourism Industry of Yogyakarta, Indonesia a Decade After the Crisis. In *Tourism, Resilience and Sustainability: Adapting to Social, Political and Economic Change*, ed. Joseph M. Cheer and Alan A. Lew, 149–164. Abingdon, Oxford: Routledge.

Davenport, A.G. 1990. The Impact of Structural Damage in Jamaica Due to Hurricane Gilbert and the Prospects for Disaster Reduction. *Journal of Wind Engineering and Industrial Aerodynamics* 36: 53–62. https://doi.org/10.1016/0167-6105(90)90292-k.

Davidson, Debra J. 2010. The Applicability of the Concept of Resilience to Social Systems: Some Sources of Optimism and Nagging Doubts. *Society & Natural Resources* 23 (12): 1135–1149. https://doi.org/10.1080/08941921003652940.

Department for International Development. 2011, November. *Defining Disaster Resilience: A DFID Approach Paper*. London.

Dodman, David, Jessica Ayers, and Saleemul Huq. 2008. Building Resilience. In *State of the World 2009: Into a Warming World*, ed. The Worldwatch Institute, 151–168. Washington, DC: Worldwatch Institute.

Doodnath, Alina. 2020, April 3. Home Gardens Take the Spotlight as Trinis Avoid Virus Spread. *Loop News*. https://www.looptt.com/content/home-gardens-take-spotlight-trinis-avoid-virus-spread.

Foley, Aileen. 2020, November 20. *SUNRISE: Researching Resilience in Islands*. Proceedings of Researching Resilience in Islands, Online. Accessed May 7, 2021. https://storymaps.arcgis.com/stories/c39f201ecb22417c9d2256adb045c352.

Folke, Carl. 2006. Resilience: The Emergence of a Perspective for Social-Ecological Systems Analyses. *Global Environmental Change* 16 (3): 253–267. https://doi.org/10.1016/j.gloenvcha.2006.04.002.

Government of the Bahamas. 2003. Ministry of Works & Utilities. *The Bahamas Building Code Third Edition—2003. By the Government of the Bahamas*. 1–935. https://www.bahamas.gov.bs/wps/wcm/connect/d7ebcbad-f9b6-42e3-aff2-79f83bd91810/Bahamas%2BBuilding%2BCode%2B3rd%2BEd.pdf?MOD=AJPERES.

Hall, C. Michael. 2018. Resilience in Tourism: Development, Theory and Application. In *Tourism, Resilience and Sustainability: Adapting to Social, Political and Economic Change*, ed. Joseph M. Cheer and Alan A. Lew, 18–33. Abingdon, Oxford: Routledge.

Hartnell, Neil. 2021, March 29. *COVID Inflicts $3bn Tourist Spending Blow*. http://www.tribune242.com/news/2021/mar/29/covid-inflicts-3bn-tourist-spending-blow/.

Holladay, Patrick J. 2018. Destination Resilience and Sustainable Tourism Development. *Tourism Review International* 22 (3): 251–261. https://doi.org/10.3727/154427218x15369305779029.

Holling, C.S. 2001. Understanding the Complexity of Economic, Ecological, and Social Systems. *Ecosystems* 4 (5): 390–405. https://doi.org/10.1007/s10021-001-0101-5.

Holling, C.S. 1973. Resilience and Stability of Ecological Systems. *Annual Review of Ecology and Systematics* 4: 1–23. https://doi.org/10.1146/annurev.es.04.110173.000245.

Ibanescu, Bogdan-Constantin, Mihail Eva, and Alexandra Gheorghiu. 2020. Questioning the Role of Tourism as an Engine for Resilience: The Role of Accessibility and Economic Performance. *Sustainability* 12 (14): 5527. https://doi.org/10.3390/su12145527.

Ilies, Gabriela. 2018. Between Resilience and Preservation Strategies: Traditional Villages from Maramures Land, Romania. In *Tourism, Resilience and Sustainability: Adapting to Social, Political and Economic Change*, ed. Joseph M. Cheer and Alan A. Lew, 295–316. Abingdon, Oxford: Routledge.

International Air Transport Association (IATA). 2020, May 5. IATA Calls for Passenger Face Covering and Crew Masks. *IATA*. https://www.iata.org/en/pressroom/pr/2020-05-05-01/.

International Monetary Fund. 2020, April 1. World Economic Outlook, April 2020: The Great Lockdown. *International Monetary Fund*. https://www.imf.org/en/Publications/WEO/Issues/2020/04/14/weo-april-2020.

IPCC. 2012. Managing the Risks of Extreme Events and Disasters to Advance Climate Change Adaptation. In *A Special Report of Working Groups I and II of the Intergovernmental Panel on Climate Change*, ed. Christopher B. Field, Vicente Barros, Thomas F. Stocker, Qin Dahe, David Jon Dokken, Kristie L. Ebi, Michael D. Mastrandrea, et al., 1–522. Cambridge, UK: Cambridge University Press.

Jouannic, Gwenaël, Anaïs Ameline, Kelly Pasquon, Oscar Navarro, Chloé Tran Duc Minh, Abdel Halim Boudoukha, Marie-Aude Corbillé, et al. 2020. Recovery of the Island of Saint Martin After Hurricane Irma: An Interdisciplinary Perspective. *Sustainability* 12 (20): 8585. https://doi.org/10.3390/su12208585.

Kopp, Ryan, Judith Mair, Terry DeLacy, and Martin Fluker. 2014. Climate Change Adaptation: Destination Management and the Green Tourist. *Tourism Planning & Development* 12 (3): 300–320. https://doi.org/10.1080/21568316.2014.988879.

Kronenberg, Roger, and Hoe Ee Khor. 2016. Economic Growth in the Pacific Island Countries—Challenges, Constraints and Policy Responses. In *Resilience and Growth in the Small States of the Pacific*, ed. Hoe Ee Khor, Roger Kronenberg, and Patrizia Tumbarello, 3–14. Washington, D.C.: International Monetary Fund.

Kruse, Sylvia, Daniel Baumgartner, Marco Pütz, and Manuela Stiffler. 2013. Vulnerability and Adaptation to Climate Change in the Alpine Space: A Case Study on the Adaptive Capacity of the Tourism Sector. In *European Climate Vulnerabilities and Adaptation: A Spatial Planning Perspective*, ed. Philipp Schmidt-Thomé and Stefan Grieving, 273–287. Chichester, West Sussex: Wiley.

La Grenade, Wendy. 2008. Balancing Economic Development and Security: Tourism and HIV/AIDS. *Social and Economic Studies* 57 (2): 27–60.

Leiper, Neil. 1979, December. The Framework of Tourism: Towards a Definition of Tourism, Tourist and the Tourist Industry. *Annals of Tourism Research* 6 (4): 390–407. https://doi.org/10.1177/004728758001900184.

Lewis, Gavin. 2020. May 24. *MOA's Kitchen Garden Project Ensuring Food Security amid COVID-19*. Department of Public Information. https://dpi.gov.gy/moas-kitchen-garden-project-ensuring-food-security-amid-covid-19/.

Manyena, Siambabala Bernard. 2006. The Concept of Resilience Revisited. *Disasters* 30 (4): 434–450. https://doi.org/10.1111/j.0361-3666.2006.00331.x.

Mendes-Franco, Janine. 2021, May 1. Heavy Rains Worsen the Effects of St. Vincent and the Grenadines' La Soufrière Volcano. *Global Voices*. https://globalvoices.org/2021/05/01/heavy-rains-worsen-the-effects-of-st-vincent-and-the-grenadines-la-soufriere-volcano/.

Mitchell, Tom, and Katie Harris. 2012 January. *Resilience: A Risk Management Approach*. Background Note. London: Overseas Development Institute.

Morton, Novelette, Devon Liburd, and Carolyn James. 2010. St. Kitts and Nevis Marketing Strategies. In *Marketing Island Destinations: Concepts and Cases*, ed. Acolla Lewis-Cameron and Sherma Roberts, 121–132. New York, NY: Routledge.

Mowforth, Martin, and Ian Munt. 2003. *Tourism and Sustainability: New Tourism in the Third World*. London: Routledge.

OCHA-UN. 2018. "Impact of Hurricanes Irma and Maria Conference Supporting Document." *Office for the Coordination of Humanitarian Affairs Assessment Capacities Project United Nations Development Programme, 2018*. https://reliefweb.int/sites/reliefweb.int/files/resources/UNDPRegionalOverviewImpactofHurricanesIrmaandMaria.pdf%0Ahttps://goo.gl/z185yj.

ODPEM. 2019. *National Structure*. Office of Disaster Preparedness and Emergency Management. https://www.odpem.org.jm/national-structure/.

Pain, Adam, and Simon Levine. 2012. *A Conceptual Analysis of Livelihoods and Resilience: Addressing the 'Insecurity of Agency.'* Humanitarian Policy Group, Overseas Development Institute, November. https://www.refworld.org/pdfid/523ac7384.pdf.

Peterson, Ryan R., Rich Harrill, and Robin B. Dipietro. 2017. Sustainability and Resilience in Caribbean Tourism Economies: A Critical Inquiry. *Tourism Analysis* 22 (3): 407–419. https://doi.org/10.3727/108354217x14955605216131.

Poon, Auliana. 1993. *Tourism, Technology and Competitive Strategies*. Wallingford, Oxfordshire: CABI.

Ritchie, B. 2017. Strategies for Tourism Business Recovery. In *Strengthening Tourism Business Resilience Against the Impact of Terrorist Attack: Report on the APEC Counter-Terrorism Working Group's Workshop*, ed. G.K. Fitriani and A. Rebekha, 61–63. Bali, Indonesia: Asia-Pacific Economic Cooperation.

Robinson, Stacy-Ann. 2020. *Knowledge, Attitudes and Practices Towards Resilience in the Caribbean*. Paper Presented at Researching Resilience in Islands, Online, November 16–17, 2020. https://storymaps.arcgis.com/stories/c39f201ecb22417c9d2256adb045c352.

Scheyvens, Regina, and Janet Momsen. 2008, December. Tourism in Small Island States: From Vulnerability to Strengths. *Journal of Sustainable Tourism* 16 (5): 491–510. https://doi.org/10.2167/jost821.0.

Schweizer, Pia-Johanna., and Ortwin Renn. 2019. Governance of Systemic Risks for Disaster Prevention and Mitigation. *Disaster Prevention and Management: An International Journal* 28 (6): 862–874. https://doi.org/10.1108/dpm-09-2019-0282.

Sgrò, Carla M., Andrew J. Lowe, and Ary A. Hoffmann. 2010. Building Evolutionary Resilience for Conserving Biodiversity Under Climate Change. *Evolutionary Applications* 4 (2): 326–337. https://doi.org/10.1111/j.1752-4571.2010.00157.x.

Sheller, Mimi. 2020. Reconstructing Tourism in the Caribbean: Connecting Pandemic Recovery, Climate Resilience and Sustainable Tourism through Mobility Justice. *Journal of Sustainable Tourism*, 1–14. https://doi.org/10.1080/09669582.2020.1791141.

Smith, Kareem. 2020, November 25. The Club Barbados Vows to Pay Axed Workers in Full. *Barbados Today*. https://barbadostoday.bb/2020/11/25/the-club-barbados-vows-to-pay-axed-workers-in-full/.

Société Internationale de Télécommunications Aéronautiques (SITA). Rep. 2021. *2020 Air Transport IT Insights*. Geneva, Switzerland: SITA.

Srinivasan, Krishna, Sònia Muñoz, and Ding. 2021, March 12. How the Caribbean Can Avoid Becoming a COVID-19 Long-Hauler. *IMF*. https://www.imf.org/en/News/Articles/2021/03/11/na031221-how-the-caribbean-can-avoid-becoming-a-covid-19-long-hauler.

Stockholm Resilience Centre. 2015. *Applying Resilience Thinking—Seven Principles for Building Resilience in Social-Ecological Systems*. Stockholm, Sweden: Stockholm Resilience Centre. https://www.stockholmresilience.org/download/18.10119fc11455d3c557d6928/1459560241272/SRC+Applying+Resilience+final.pdf.

The Story Editor. 2021, February 20. Backyard Gardening Eases, Minimizes COVID-19's Negative Effects on Families. *The St Kitts Nevis Observer*. https://www.thestkittsnevisobserver.com/backyard-gardening-eases-minimizes-covid-19s-negative-effects-on-families/.

The World Bank Group. 2016, February 18. *The Short-Term Economic Costs of Zika in Latin America and the Caribbean (LCR)*. Washington: The World Bank Group.

Thompson, Peter. 2020, April 28. *Barbados Beyond Tourism*. Wunnuh Consultants. http://wunnuh.org/barbados-beyond-tourism/.

Trocaire. 2012, January 11. *Building a new Haiti from the rubble of traged*. https://www.trocaire.org/news/building-a-new-haiti-from-the-rubble-of-tragedy/.

United Nations Environment Programme. Rep. 1989. *Assessment of the Economic Impacts of Hurricane Gilbert on Coastal and Marine Resources in Jamaica*. Cep Technical Report No. 4. Kingston, Jamaica: UN Environment Programme.

United Nations Office for Disaster Risk Reduction (UNDRR). Rep. 2012. *Making Cities Resilient Report 2012*. Geneva, Switzerland: UNDRR. https://www.unisdr.org/files/28240_rcreport.pdf.

Wallemacq, Pascaline and Rowena House. 2018, October. *Economic Losses, Poverty & Disasters 1998–2017*. Cred Crunch, No. 52. https://www.undrr.org/publication/economic-losses-poverty-disasters-1998-2017.

Wilkinson, Emily, and Katie Peters. 2015. *Climate Extremes and Resilient Poverty Reduction: Development Designed with Uncertainty in Mind*. London: Overseas Development Institute.

World Tourism Organization. 2020, March 24. Impact Assessment of the COVID-19 Outbreak on International Tourism. *UNWTO*. https://www.unwto.org/impact-assessment-of-the-covid-19-outbreak-on-international-tourism.

Worrell, DeLisle. 2010. *The Barbados Economy in 2009 and Prospects for 2010*. Bridgetown. Barbados: Central Bank of Barbados. https://www.bis.org/review/r100211c.pdf.

Wu, Emily, and Geoffrey Wall. 2018. Learning from Dabang, Taiwan: Sustainability and Resilience in Action in Indigenous Tourism Development. In *Tourism, Resilience and Sustainability: Adapting to Social, Political and Economic Change*, ed. Joseph M. Cheer and Alan A. Lew, 222–242. Abingdon, Oxford: Routledge.

PART II

Environmental Resilience

CHAPTER 4

Implications for Resilience in the Cruise Tourism-Marine Protected Area Nexus in the Caribbean: The Case of St. Lucia

Myrna Ellis

Introduction

The cruise tourism industry developed in stages with the very first ocean pleasure cruise occurring in 1881 (Raluca and Monica 2008). By the 1920's cruising became the most favoured means of travel by the elite. Patullo (1996) noted that the sea has always been an economic highway for slavers, traders, buccaneers and fishermen. Today the Caribbean Sea acts as an economic highway that brings cruise visitors to the region. Raluca and Monica (2008, 630) defined a cruise as "*a multi-centre holiday where you take your hotel with you from place to place*". It is characterized by ships which are comparable to moving resorts, transporting passengers from place to place (Dowling 2006).

Often described as a cultural, social and economic unit rich in biodiversity, the Caribbean attracts many cruise ships to its shores. Two factors

M. Ellis (✉)
Caribbean Maritime University, Kingston, Jamaica

A. Lewis-Cameron et al. (eds.), *Managing Crises in Tourism*,
https://doi.org/10.1007/978-3-030-80238-7_4

attributed to this phenomenon are its tropical climate and its rich biodiversity. The Caribbean is also one of many "biodiversity hotspots" in the world. Johnson (2002) and Twining-Ward (1999) noted that cruise passengers are invariably attracted to "biological hotspots". Approximately 70% of cruise destinations can be found in hotspots like the Caribbean (Sweeting & Wayne 2003). Consequently, destinations located in these hotspots struggle to absorb the additional pressures cruising exerts on their fragile ecosystems (Johnson 2002). These added stressors impact resilience in marine spaces.

The CT-MPA nexus is an area in which both conservation and recreational use (by multiple stakeholders) occur simultaneously. Marine protected areas (MPAs) are management tools aimed at reducing biodiversity loss to ensure that marine ecosystems continue to provide essential ecosystem services including food such as fish (the primary source of protein globally) and recreation such as tourism. Lately, it appears that the ecosystem services provided by MPAs are competing with each other, given that the recreational services provided through cruise tourism appear to threaten provisional services including food, water quality and pest control. This has implications for resilience of the CT-MPA nexus in the short term as well as the well-being of surrounding communities in the long term.

This research assesses MPAs as an approach to improve resilience. It identifies physical-ecological carrying capacity indicators (EC 2000) used to gain insight into whether cruise tourism has triggered a decline in ecosystem resilience on account of compromised physical-ecological carrying capacity levels (Coccossis and Mexa 2004). The aim of this research is to: (i) identify the challenges presented by cruise tourism; (ii) evaluate the measures taken to enhance resilience; (iii) discuss lessons learnt and (iv) propose recommendations for enhancing resilience.

Literature Review

There is no one precise definition of the Caribbean rather, its definition depends on context. In this research, "Caribbean" refers to the English-speaking islands of CARICOM (CARSEA 2007). The region is four times more tourism-dependent than any other region in the world and its cruise sector has experienced consistent growth. In 2018, cruise ships brought 15.44 million passengers to the region, representing an economic impact of US$1.48 billion. (FCCA 2018). Additionally, cruise related expenditures generated 45,225 jobs and paid US $728.1 million in wages to employees in the industry (FCCA 2018).

The Tourism-Environment Relationship

From inception, taking pleasure in environments has had a major role in making tourism what it is today. However, the relationship between tourism and the environment is both multifaceted and symbiotic. Given that tourism profits from being in environments of a high quality, the tourism industry should institute measures to protect these environments (Williams 1998). With the rapid growth of tourism however, it has become indisputable that the symbiosis is now unbalanced and tourism is now the source of serious environmental issues. Consequently, efforts are being made firstly, to understand the impacts of tourism on the environment and secondly, to generate forms of tourism that sustain rather than degrade key resources. Given the varied and diverse nature of these impacts, disentangling tourism's influences from other agents of change remains one of the many complications of the tourism-environment relationship (Williams 1998).

Research suggests that environmental stability in small islands has been compromised by the practice of mass tourism and has disturbed the natural peace of island life (Briguglio and Briguglio 2002). Coastlines have become modified and endemic species have been reduced in number due to construction of large resorts and marinas and other infrastructural work. In addition, tourist activities may result in the corruption of land, ocean and coastal areas with solid and liquid waste produced by hotels, marinas and cruise ships (Neto 2003). This has become a real issue for small island systems that lack the requisite assimilative capacity, and are characterized by scarce resources, immense vulnerability, human resource constraints, limited expertise, burdensome bureaucracy, overlapping jurisdictions and the general lack of interagency co-operation (Neto 2003).

One of the most attractive settings for tourism has always been the marine environment. Marine tourism is defined as: "*those recreational activities that involve travel away from one's place of residence and which have the marine environment i.e. those waters which are saline and tide-affected as their focus*" (Orams 1999, 9). These recreational activities he noted include: fishing, scuba diving, snorkeling, the cruise and ferry industry and beach activities such as kayaking, skiing, sailing and yachting and many more. MPAs play an important role as a resilience tool for safeguarding marine ecosystems and coastal communities. As a result of a call by the International Union for the Conservation of Nature to protect 10% of marine waters by 2020, there has been a noticeable increase in the number of MPAs globally. Kelleher (1999, xviii) defines a MPA as "*a clearly defined geographic space recognized, dedicated and managed, through legal or other effective means to achieve the long-term conservation of nature with associated ecosystem services and cultural values*". The CT-MPA nexus is an area designated as a MPA in which cruise visitors participate in many of the abovementioned recreational activities. According to Spalburg (2009), 1.6% of all tourists participate in the cruise sector. For such a small niche which is reliant on the marine environment, its impacts are disproportionate to its size presenting challenges which are rooted in the concept of sustainable tourism (Farrell & Twinning-Ward 2004).

A Call for Sustainable Tourism

For countries like St. Lucia, whose economy is driven by tourism, the environmental problems associated with unsustainable tourism are of immense concern. Neto (2003) proposed promoting sustainable tourism in order to reduce tourism's environmental impact and take advantage of its benefits. Carter (1993) identified three key objectives of sustainable tourism: (i) improving the standard of living of the host population in the short and long term; (ii) satisfying the demands of growing numbers of tourists and (iii) safeguarding the natural environment in order to achieve both of the preceding goals. The UNEP &WTO (2005) defined sustainable tourism as: *"tourism that takes full account of its current and future economic, social and environmental impacts while addressing the needs of visitors, the industry, the environment and the host community"*.

One tool for measuring sustainable tourism is that of carrying capacity. Butler (1980) established the conceptual basis of carrying capacity in tourism planning in his Tourism Area Life Cycle (TALC). The TALC concept proposes that tourism cannot grow continuously in a place without causing irreversible damage to the local system. It conveys the idea of limits to growth. Mc Cool and Lime (2001) argue that the primary question underlying carrying capacity should not be "how much is too much" but rather "how many changes to the environmental conditions are acceptable" The problem however, is that development efforts tend to be cumulative, making it difficult if not impossible to forecast the final impact that many incremental changes can have over time.

Symmonds and Hammitt (2000) and Williams (1998) proposed three types of parameters for determining tourism carrying capacity: physical-ecological; socio-cultural and political-economic. The physical-ecological parameters refer to fixed and flexible components. The fixed components (ecological capacity) refer to the assimilative capacity that allows the ecosystems and resources to absorb loads of pollutants, emissions etc. without the loss of its function while the flexible components (physical capacity) refer to elements of the built environment (EC 2000).

Cruise tourism can have both direct and indirect effects on the physical-ecological aspect of carrying capacity and hence the resilience of MPAs, given that cruise ships produce a variety of waste streams and cruise visitors participate in various activities. Calculations have revealed that a cruise ship with a capacity of some 2,000–3,000 passengers can generate 1,000 tons of waste per day. Food waste contributes to the increase in biological oxygen demand diminishing water quality and harming fish (Polgaze 2003) while plastic waste poses a threat to sea life. The intense use of tourism-related water-based transportation contributes to noise pollution, an underestimated pollutant, while fuel from these vessels increases air pollution. In addition, activities such as snorkelling, scuba diving and speed boats can disrupt marine life having a negative effect on fisheries (Neto 2003). By far, the most serious threat to coral reefs is anchor damage which is driven by the increased number of vessels traversing the Caribbean Sea (Hall 2001). The future of the local tourism and fishing industry may be at risk if coral reef damage is allowed to continue unabated (Neto 2003).

Wilkinson (2006) cautioned that Caribbean tourism cannot continue to grow at its present rate given that this type of activity increases the number, type and severity of impacts to marine ecosystems. The cruise industry relies on these ecosystems for its existence and Wilkinson (2006) further suggested that focus be placed on determining the carrying capacity of islands for this specific activity. For ecosystem services to be sustained overtime, the ecosystems providing them must be able to continue functioning in essential ways despite disruptions. They must be resilient i.e. have the capacity to keep functioning even when disturbed or have the ability to recover from disturbances. Disruption of marine ecosystems diminishes ecosystem services such as: the provision of fish and other seafood, the maintenance of water quality and the control of pests and pathogens. It also reduces resilience.

Methodology

The Study Area

St. Lucia lies between 60' and 61' west longitude and 13' and 14' north latitude and is renowned for its safe and strategically located harbour. The island is part of the Windward Island chain in a southern group of islands in the Lesser Antilles in the West Indies and has an area of 239 square miles (616 square kilometers). It has a population of 172,811 and its official language is English yet many people speak Patois, a type of French. 40% of the population live in Castries, the capital, while 15.2% live in Gros Ilet (St. Lucia Central Statistical Office 2010). St. Lucia's cruise sector has been growing steadily and the destination hosted over 600,000 cruise visitors in 2019 and visitor expenditure accounted for over 55% of the country's GDP. Tourists arrive in St. Lucia via airlift, cruise ships and yachts. The cruise season peaks between December and March each year and accounts for more arrivals than stay over visitors. During these months, cruise visitors disembark at one of two private modern duty-free areas (Pointe Seraphine or La Place Carenage), and are taken by water or land taxi to popular sites and attractions as shown in Fig. 4.1. One such attraction is the dive site in CT-MPA nexus in Soufriere, a small fishing village located in the south of the island as shown in Fig. 4.2.

To realize the objectives of the research it was necessary to gather information from persons both directly and indirectly involved in cruise tourism in the marine environment. For this reason, a mixed method approach was used in which repeat cruise visitors, tour operators/guides, fishermen and experts from the cruise industry (seen in Table 4.1) were chosen as Interviewees.

Approximately 123,000 arrived by cruise ships in the month of January, and 17,729 disembarked from 13 ships. Using a confidence level of 95% a sample size of 377 was required (Raosoft 2004). The research employed a mixed method concurrent triangulation in which qualitative and quantitative data were collected concurrently due to time constraints, limited resources and availability of personnel. Data were collected with the assistance of 20 undergraduate tourism students who were trained and paid 3USD per questionnaire. Qualitative data were collected from

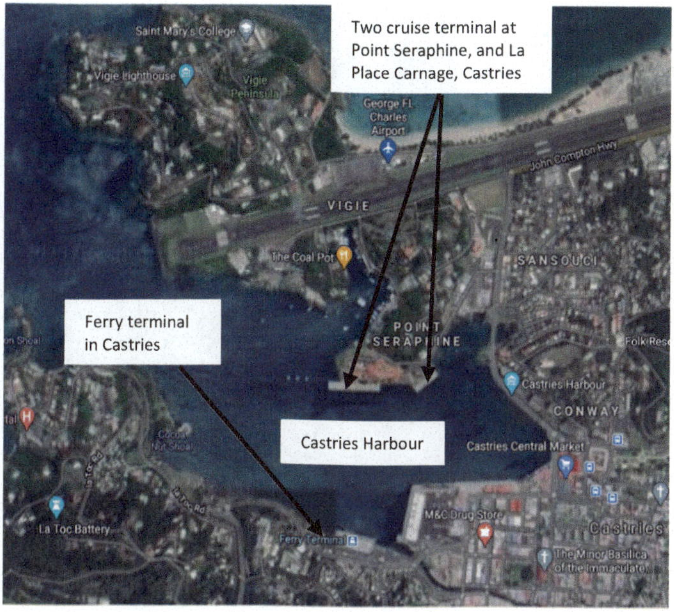

Fig. 4.1 Cruise Terminals in Castries, St. Lucia

8 elite interviewees using semi-structured interviews and 2 focus groups comprising 10 fisher folk each from Castries and Soufriere while quantitative data were collected via 284 surveys from cruise visitors and tour operators. The qualitative and quantitative data were analysed using content analysis and SPSS respectively, and the results were triangulated to elicit meaning as shown in Fig. 4.3.

Findings & Discussion

The quantitative and qualitative findings revealed that the demands of a fast-growing cruise industry in St. Lucia may be contributing to several of the changes observed in the marine environment and hence may have an effect on the environment's ability to provide needed ecosystem services. If these services can no longer be provided, human well-being may be at risk (MEA 2003, 2005). These changes include: (i) polluted and

Fig. 4.2 The CT-MPA nexus in Soufriere

Table 4.1 Description of elite interviewees

Interviewee	*Description*
Interviewee #1	Director of the St. Lucia Sustainable Tourism Development Programme
Interviewee #2	Fisheries Biologist; Former General Manager Soufriere Marine Management Authority
Interviewee #3	Senior Officer, Ministry of Sustainable Development, Energy, Science and Technology
Interviewee #4	Senior Manager, St. Lucia Air & Seaport Authority (SLASPA)
Interviewee #5	Operations Manager at a major tour company in St. Lucia
Interviewee #6	Dive shop owner; Diver
Interviewee #7	Senior Ranger, Soufriere Marine Management Authority
Interviewee #8	Hotel Manager, Castries

discoloured water; (ii) damaged reefs; (iii) decreased fish populations; (iv) increased solid waste and (v) increased conflict among users.

Polluted and discoloured water: When asked about recreational water quality, most (seven out of eight) of the elite interviewees had

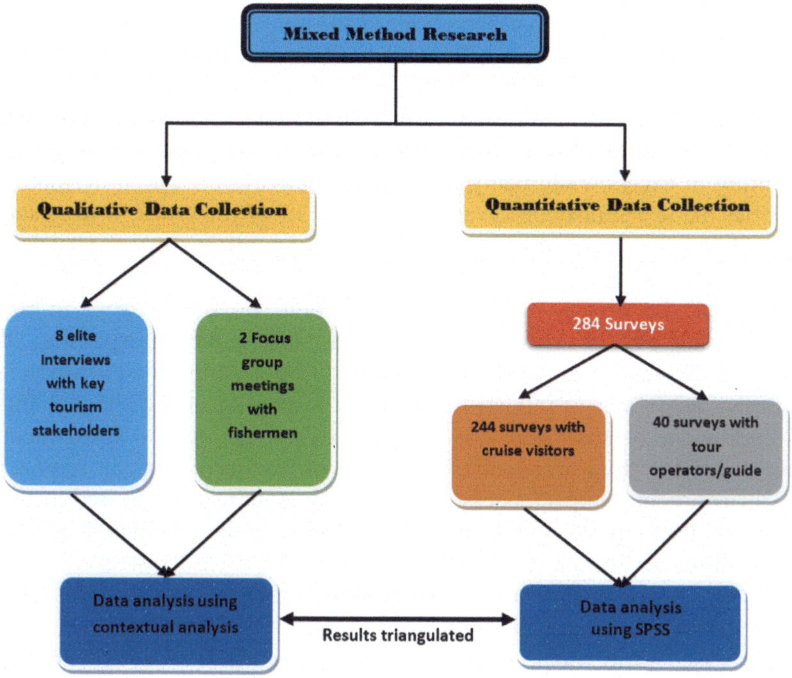

Fig. 4.3 Mixed Method Research Design

no issues with the quality of the water in St. Lucia and stated that the Department of Fisheries conducted water quality testing regularly. The quantitative data showed that 90.4% of the cruise visitors indicated that they had no complaints in this regard however, 10% of the tour operators stated that their guests had noticed changes in the quality of the water.

When asked about the discolouration of the water however, interviewees #3 and #8 lamented the discolouration of water in the Castries harbour over time. This was reiterated by both focus groups who further added that the water was not only discoloured but sometimes had an oily substance on its surface. Neto (2003) noted that high levels of fuels used by tourism-related transportation contribute to the pollution of waterways. Additionally, the quantitative data revealed that 16.7% of the cruise visitors and 45% of the tour operators noticed discoloured water.

Interviewees were also asked about cruise ships dumping waste at sea. Interviewee #8 stated that effluent had been dumped in the Castries harbour and suggested that this may be a contributor to the discolouration seen. Another interviewee added that some catamarans can only support 100 toilet flushes but at times exceeded the carrying capacity resulting in excess waste being dumped in the harbour. The fisherfolk in Castries made a similar observation while 26.4% of the cruise visitors and 25% of the tour operators stated that they had either seen or heard of cruise ships dumping waste at sea.

One of the drivers of these changes could be the increase in the number and size of cruise ships visiting St. Lucia due to expansion of the port. Presently, as many as 6 vessels can be accommodated simultaneously. As a result of infrastructural work (dredging of the channel and extending the berths) there was increased sediment load and consequent loss of marine flora and fauna. This type of expansion usually requires an environmental impact assessment (EIA) be conducted as stated in the Barbados Programme of Action (1994). However, the St. Lucia National Environmental Summary (2010) indicated that EIA legislation needs revamping to consider stronger post EIA monitoring and a stronger focus on strategic environmental assessment.

Increased solid waste: Interviewee #6 revealed that the reefs were full of solid waste while interviewee #8 added that the beaches were polluted with solid waste which usually occurred on public holidays when cruise visitors crowd-out the stay-over visitors at the beaches. She stated that visitors had complained about excessive amounts of solid waste at Rodney Bay, especially on public holidays when cruise ships were in port. Additionally, the fishermen noticed an increase in both the dumping of waste at sea and plastic material at reef sites. From the quantitative findings, 16.7% of the cruise visitors and 20% of the tour operators noticed floating objects in the sea. According to Cheshire and Alder (2009), increases in cruise visitor arrivals results in increased solid waste and the need for efficient solid waste management.

Neto (2003) pointed out that tourist activities can lead to contamination of land and marine resources with solid waste, a problem exacerbated by the rapid growth of cruise tourism. Beaches that are polluted when cruise ships dock on public holidays indicate that overcrowding may be an issue, as locals also add to the already high numbers at the beach. This raises the question of whether carrying capacity levels may be compromised on occasions such as these. One of the interviewees, a diver, pointed out that he has participated in many reef clean-up exercises in which large amounts of plastic waste was removed from the reef. Cheshire and Alder (2009) noted this to be an adverse effect of cruise tourism.

Damaged reefs: Interviewee #1 commented on the damaged state of the reefs in St. Lucia, a situation which he stated accelerated over time. This was confirmed by the fishermen and cruise passengers. Interviewee #1 attributed this damage to overcrowding at the reefs and this was supported by interviewee #7 who stated that one reef site was sometimes visited by over 200 snorkelers simultaneously. Interviewee #2, concurred with this, pointing out that on some days when there are 5–6 cruise ships in port, as many as 2000–3000 cruise visitors are taken by day-boat charters to the reefs.

Anchorage of vessels, driven by an increase in the number of dive boats and catamarans anchoring near the reef, (Rogers et al. 1988) especially in Jalousie, may be a source of reef damage. Furthermore, interviewee #1 stated that when 200 inexperienced snorkelers and divers along with 2000–3000 cruise passengers overcrowd the reef, this causes damage, suggesting that permissable tourism carrying capacity levels may have been overstepped.

The focus group in Soufriere also indicated that in addition to climate change, the corals are under stress from sediment build up in the water and algal bloom from land- based activities. The sediment plumes seen at a dive site in Jalousie, resulted from the importation of white sand from Trinidad which was used for beach rehabilitation along with the construction of hotels along the coastline with no regard for setback distances (Mc Elroy and Albuquerque 1998).

Decrease in fish population: Four out of eight interviewees indicated that guests to St. Lucia noticed less fish at the reef while interviewee #8 stated that less fish was being caught. The fishermen also noted that their daily fish catch was much less than usual and further added that this decrease occurred concomitantly with the growth of cruise tourism. Furthermore, interviewee #8 revealed that cruise visitors complained about going to the reef and seeing less and less fish. Cohen (2006) posited that the expansion of coastal tourism activities may cause a decrease in fish populations.

Five out of eight elite interviewees alluded to increasing numbers of vessels, driven by the growth in unregulated and unregistered vessels, as reasons for the decrease. Additionally, the unauthorized use of fishing priority areas by dive-boats operators and their guests disturbed and destroyed fish traps and equipment set by fishermen, driving fish away, causing conflict. Even cruise ships cross the path of fish traps and equipment, causing disturbance. Rogers et al. (1988) stated that ship engines are the main source of noise pollution which significantly impacts the marine environment negatively due to noise amplification in water. A study by Cohen (2006), revealed that a ten-fold increase in cruise related traffic in Yakutat Bay, Croatia caused a 66% reduction in seal population in the Bay over a ten year period.

In-depth analysis of the qualitative data revealed that 80% of the elite interviewees and fishermen alluded to: "*increase in informal sector*", "*undocumented tours*", "*increased noise pollution*", "*water pollution by oil and diesel*" suggesting concerns over the increase in small vessels traversing the marine space while 80% also referred to: "*overcrowding of beaches*", "*overuse of dive sites*", "*overfishing*" expressing concerns with the exceeding of carrying capacity levels. Additionally, when asked about the management of the marine environment, 90% indicated that the establishment of the Soufriere Marine Management Authority (SMMA) has made a significant difference in this regard and made reference to: "*SMMA model*", "*zoning*", "*demarcation of boundaries*", "*monitoring*", "*conflict resolution*" but further stated that the effectiveness of the model was severely hampered by the lack of enforcement of regulations using expressions such as: "*overlapping jurisdictions*", "*outdated laws*", "*lenient fines*", "*lack of political will*", "*non-payment of fees*".

> Overall, the findings show that the demands of the growing cruise industry may be contributing to the changes in the CT-MPA nexus. This challenges resilience in the marine environment given that the environment's ability the ecosystem services necessary for human well-being may be compromised (MEA 2003).

Management Strategies: Interviewees were asked to identify the measures perceived to be the most effective management strategy for minimizing the threat to the physical-ecological carrying capacity level and enhance resilience in the marine environment. The results revealed that enforcement of regulations was perceived to be the most effective management strategy, while an environmental fee was least effective. These findings concur with Orams (1999) who proposed education and enforcement as effective management strategies but contradict his statement that an environmental fee is also effective.

Enforcement of regulations: All interviewees indicated that environmental regulations are not enforced in St. Lucia. Interviewees of the focus groups stated that "many fishermen break the law because they have families to provide for". Poor coordination, weak and inadequate existing laws and weak enforcement of current environmental regulations are some of the gaps hindering appropriate resource management (National Environmental Summary 2010).

Education and awareness: 81.2% of cruise visitors and 50% of tour operators were unaware of environment regulations in the marine environment. In addition, 30% of the tour operators stated that while conducting their tours, they did not engage visitors in discussions about their role in protecting the marine environment. Education can be a valuable tool in increasing awareness and Orams (1999) proposed that destinations incorporate environmental messages as an educational management strategy.

Environmental fee: The quantitative and the qualitative findings both suggest that an environmental fee is the least effective strategy despite the fact that 78.2% of cruise visitors were willing to pay a fee of at least 5 USD. interviewees stated:

- *"Most cruise visitors pay the environmental fee and do not know what it is for".*
- *"Boat operators using the Marine reserve in Soufriere collect the fee from visitors but it is not remitted to the SMMA".*
- *"Most boat operators are unwilling to pay the fee because they believe that they receive no benefits from it".*

Implications for resilience in the CT-MPA

The foregoing challenges may have far-reaching implications for the resilience of the CT-MPA Nexus. For ecosystem services to be sustained overtime, the ecosystems providing them must be able to continue functioning despite disruptions. Documents such as the Barbados Programme of Action (1994), the Convention on Biological Diversity (2002) and the Barbados Tourism Policy (2000) along with writers such as Orams (1999) and Wilkinson (2006) have all emphasized the importance of determining tourism carrying capacity especially in marine environments. Using the capacity levels and indicators proposed by Coccossis and Mexa (2004) and the challenges revealed by the research findings, Table 4.2 shows the level of threat to the carrying capacity threshold and suggests that the physical ecological TCC levels may be compromised thus decreasing resilience in the CT-MPA nexus.

Enhancing resilience in the CT-MPA nexus

Currently, St. Lucia hosts over 600,000 cruise visitors annually (SLASPA 2014). According to Weaver (1993), such occurrences affect the lives of people living near the port on a daily basis. Furthermore, despite being rich in natural resources and cultural assets, St. Lucia faces unique challenges with respect to planning and sustainable development. Notwithstanding, several initiatives have been undertaken in an effort to overcome these challenges. Firstly, St. Lucia is signatory to several international conventions regarding the marine environment and a national regulatory framework exists to guide St. Lucia with respect to the marine environment. Secondly, the Sustainable Development and Environment Unit of the Ministry of Planning, was established in 2001 with responsibility for coordinating national planning for sustainable development including coastal and marine, freshwater and land resources which are critical to sustainable development in St. Lucia (St. Lucia National Report 2001). Thirdly, recreational water quality standards were developed in 2009 to facilitate the implementation of the land-based sources protocol and in collaboration with Caribbean Environmental Health Institute, a pilot project was launched in 2010 to test the effectiveness of the water quality standards (National Environmental Summary 2010).

Fourthly, the Department of Fisheries was appointed the lead agency with responsibility for the management of St. Lucia's coastal and marine resources. As the lead agency, the Department established the Soufriere Marine Management Area (SMMA). Fifthly, the SMMA Inc. an independent self-sustained not-for-profit, non-governmental organization was established by the Government of St. Lucia to manage the SMMA. The management approach used is that of co-management. Its objectives are to: (i) conserve the coastal and marine resource base of Soufriere; (ii) enhance the equitable economic, social and cultural benefits generated from the sustainable use of the coastal and marine resources of Soufriere at the local and national levels; and (iii) manage the conflicts that may occur among users of the coastal and marine resources.

Table 4.2 Perceived level of threat to the physical-ecological capacity in the CT-MPA nexus

Capacity levels for physical-ecological carrying capacity (EC 2000)	*Question to be answered*	*Indicators*	*Evidence from the research findings*	*Threat to physical-ecological capacity level*	*Impact on resilience*
Acceptable levels of congestion or density of key areas e.g. coral reefs	Is the level of congestion of coral reefs under threat of exceeding the acceptable level?	Damage to coral reefs	Anchor damage due to increased numbers of vessels; reef touching due to overcrowding (>200 snorkelers per site; > 29,000 reef touches per year)	High	Decreased resilience
Maximum acceptable loss of natural resources without significant degradation of ecosystem functions and biodiversity loss	Are natural resources under threat of exceeding the maximum acceptable loss?	Loss of marine flora and fauna (e.g. fishes)	Dredging of channel and port expansion in Castries; Increased noise pollution and turbulence from boat engines; oil slicks on the water; reduction in fish population	High	Decreased resilience
Acceptable levels of water and noise pollution on the basis of tolerance or the assimilative capacity of local ecosystems	Is the level of noise and water pollution under threat of exceeding acceptable levels?	Water quality	Increased number of small vessels; increased noise pollution and turbulence from boat engines; oil slicks on the water; Release of human waste from small vessels	High	Decreased resilience

(continued)

Table 4.2 (continued)

Capacity levels for physical-ecological carrying capacity (EC 2000)	*Question to be answered*	*Indicators*	*Evidence from the research findings*	*Threat to physical-ecological capacity level*	*Impact on resilience*
Intensity of use of transport in the marine environment, infrastructure and facilities	Is the intensity of use of transport in the marine environment exceeding acceptable levels?	No. of accidents/incidents in the marine environment	Increase in the number of licensed and unlicensed small vessels; increase in conflict among users; harassment of guests on the beach; increase in crime against visitors	High	Decreased resilience
Use and congestion of utility facilities and services	Are facilities and services under threat of congestion and overcrowding	Management and disposal of solid waste	Increased littering at popular beaches; increase in solid waste (plastic) found at coral reef sites	High	Decreased resilience

Source Adapted from the EC (2000) and Coccossis and Mexa (2004)

In this regard, SMMA Inc. instituted a number of measures aimed at enhancing resilience in the CT-MPA Nexus including: (i) zoning of the marine protected area; (ii) monitoring of coral reefs and water quality; (iii) developing schedules to regulate the use of reef sites; (iv) collecting of user fees for use of the marine park; (v) staff training; (vi) using alternative dispute resolution to manage conflict. Orams (1999) contended that, "marine park regulations" and "zoning" should be instituted as a strategy for boosting resilience in marine ecosystems especially in islands where the number of tourists on day-visits may exceed the population of the island.

Lessons Learnt & Recommendations

Several lessons emerge from St. Lucia's attempt to enhance resilience in the marine environment. They all point to the fact that there is no one-off fix to enhance resilience rather, this must be an ongoing cyclical process involving planning, doing, evaluating and taking action. St. Lucia is signatory to several international agreements. The signing of international agreements and the development of a national regulatory framework does not guarantee that regulations and laws will be adhered to. Instead, regulatory frameworks are only effective if they are enforced. Enforcement of regulations has proven to be a challenge in St. Lucia and Neto (2003) noted that lack of enforcement hinders the planning process. Additionally, many unregistered vessels continue to use the marine park on account of inadequate enforcement of regulations which has implications for carrying capacity thresholds and resilience. Orams (1999) stated that the specific capacity for which a vessel is licensed should be adhered in order to limit the numbers at a site at any point in time. Unquestionably, vessel registration and use in the marine park needs to be regulated.

Another lesson came from the remittance of user fees. The stipulated portion of the user fees collected by tour operators from the cruise visitors is seldom ever collected by the SMMA. These remittances pay salaries, purchase equipment, administer training programmes etc. One elite interviewee stated that should these remittances be collected as stipulated, the SMMA would realize over 1 million USD per year to carry out their responsibilities. Uncollected user fees add no value.

Schedules were developed for some but not all snorkelling sites. Orams (1999) noted that schedules help to vary the itinerary so as to avoid exceeding the carrying capacity and reducing resilience. As such, the integrity of some sites remains compromised. Additionally, the adherence to the schedule requires the necessary personnel for monitoring so as to avoid overcrowding. Park rangers monitor the marine park between the hours of 8:00 a.m. and 4:00 p.m. only. On the contrary, many issues and conflicts involving yachts and fishermen occur outside of those hours and remain unresolved. Incidents of theft onboard yachts and disregard for zoning occur mainly after dark. Marine parks therefore need a system of 24-h monitoring strategy which requires additional financial and human resources. Lastly, notwithstanding an initial reduction in conflicts when the SMMA was established, it was not long before the SMMA became inundated with a resurgence of disputes and conflicts between fishermen and yachters, divers and fishermen and tourism stakeholders and fishermen. This illustrates that establishing a MPA as a management tool is not a panacea for all problems.

Recent research indicates that the establishment of MPAs can build resilience against external pressures. However, for the most well-intentioned initiatives such as the SMMA, establishing the MPA alone may be insufficient. Many researchers have pointed out the complexity of these types of socio-ecological systems and call for new resilience-building management policies which are flexible and open to learning through monitoring and evaluation. Therefore, adaptive co-management, a management approach in which institutional arrangements and ecological knowledge are tested and revised in a dynamic, ongoing, self-organized process of learning-by-doing (Olssen and Folkes 2004) is strongly recommended for the CT-MPA nexus.

Conclusion

Cruise tourism continues to grow in St. Lucia and the CT-MPA nexus, an area in which conservation and use are two competing forces in a complex web of actors, continues to be impacted. The findings suggest that the physical-ecological carrying capacity levels are being threatened by increased activity in the area, thereby reducing resilience. While establishing MPAs as a management approach has been shown to reduce biodiversity loss and improve resilience, it is not a one-time fix but rather this management approach needs to be adaptive to deal with the dynamic nature of this type of ecosystem in which changes are sudden, unpredictable and cumulative. CT-MPA nexuses such as the area managed by the SMMA, should adopt adaptive co-management as the best approach for reaping socio-economic benefits (through tourism) and ecological benefits (improving resilience).

References

Briguglio, L., and Marie Briguglio. 2002. Sustainable Tourism in Small Islands: The Case of Malta. In *Tourism, Biodiversity and Information* edited by Castri and Balaji, 169–184. Leiden: Backhyus Publishers.

Butler, Richard. 1980. The Concept of a Tourist Area Cycle of Evolution. *Canadian Geographer* 24 (1): 5–12. https://doi.org/10.1111/j.1541-0064.1980.tb00970.x.

CARSEA (Caribbean Sea Ecosystem Assessment). 2007. *A Sub-Global Component of the Millennium Ecosystem Assessment,* edited by John Agard, Angela Cropper and Keisha Garcia. Caribbean Marine Studies, Special Edition.

Cater, Erlet. 1993. Ecotourism in the Third World: Problems for Sustainable Tourism Development. *Tourism Management* 14 (2): 85–90. https://doi.org/10.1016/0261-5177(93)90040-R.

Cheshire, A.C., E. Adler, J. Barbière, Y. Cohen, S. Evans, S. Jarayabhand, L. Jeftic, R.T. Jung, S. Kinsey, E.T. Kusui, I. Lavine, P. Manyara, L. Oosterbaan, M.A. Pereira, S. Sheavly, A. Tkalin, S. Varadarajan, B. Wenneker, and G. Westphalen. 2009. *UNEP/IOC Guidelines on Survey and Monitoring of Marine Litter.* UNEP Regional Seas Reports and Studies, No. 186; IOC Technical Series No. 83: xii + 120 pp. https://doi.org/10.25607/OBP-726.

Coccossis, Harry, and Alexandra Mexa. 2004. *The Challenge of Tourism Carrying Capacity Assessment: Theory and Practice.* Basingstoke, Hampshire: Ashgate.

Cohen, G. 2006. *Campaign to Safeguard Americas Waters: Overview of Cruise Ship Facts, Criminal History, Regulatory Status and Threats to the Environment (C-SAW).* Earth Island Institute.

CSO (Central Statistical Office). 2010. Population and Housing Census. Castries, St. Lucia.

Dowling, Ross. 2006. The Cruising Industry. In *Cruise Ship Tourism,* edited by Ross Dowling, 3–17. Wallingford, UK: CAB International. https://doi.org/10.1079/9781780646084.0000.

EC (European Community). 2000. *Towards Quality Coastal Tourism: Integrated Quality Management of Coastal Tourist Destinations*, 156. Luxemburg: Office for Official Publications of the European Communities.

FCCA (Florida-Caribbean Cruise Association). 2018. *Cruise Industry Overview*: Florida-Caribbean Cruise Association.

Farrell, Brian, and Louise Twining-Ward. 2004. Reconceptualising Tourism. *Annals of Tourism Research* 31 (2): 274–295. https://doi.org/10.1016/j.annals.2003.12.002.

Hall, Michael. 2001. Trends in Ocean and Coastal Tourism: The End of the Last Frontier? *Ocean and Coastal Management* 44 (9): 601–618. https://doi.org/10.1016/S0964-5691(01)00071-0.

Johnson, David. 2002. Environmentally Sustainable Cruise Tourism: A Reality Check. *Marine Policy* 26: 261–270. https://doi.org/10.1016/S0308-597X(02)00008-8.

Kelleher, G. 1999. *Guidelines for Marine Protected Areas.* Gland, Switzerland and Cambridge, UK: IUCN.

McElroy, Jerome, and Klaus de Albuquerque. 1998. Tourism Penetration Index in Small Caribbean Islands. *Annals of Tourism Research* 25 (1): 445–468. https://doi.org/10.1016/S0160-7383(97)00068-6.

McCool, Stephen F., and David Lime. 2001. Tourism Carrying Capacity: Tempting Fantasy or Useful Reality? *Journal of Sustainable Tourism* 9 (5): 372–387. https://doi.org/10.1080/09669580108667409.

MEA (Millennium Ecosystem Assessment). 2003. *Ecosystems and Human Well-Being: A Framework for Assessment*. Island Press, Washington, DC.

Neto, Fredrico. 2003. A New Approach to Sustainable Tourism Development: Moving Beyond Environmental Protection. *Natural Resources Forum*

Olsson, P., C. Folke, and F. Berkes. 2004. Adaptive Co-management for Building Resilience in Social-Ecological Systems. *Environmental Management* 34: 75–90. https://doi.org/10.1007/s00267-003-0101-7.

Orams, Mark. 1999. *Marine Tourism: Development Impacts and Management*. London: Routledge.

Patullo, Polly. 1996. *Last Resorts: The Cost of Tourism in the Caribbean*. London: Cassell.

Polglaze, John. 2003. Can We Always Ignore Ship-Generated Food Waste. *Marine Pollution Bulletin* 46 (1): 33–38. https://doi.org/10.1016/s0025-326x(02)00324-7.

Rogers, Caroline, Larry McLain, and Evonne Zullo. 1988. Damage to Coral Reefs in Virgin Islands National Park and Biosphere Reserve from Recreational Activities. *Proceedings of the 6th International Coral Reef Symposium*, Townsville, Australia 2: 405–410, 8–12 August.

Raluca, D., and Monica, G. 2008. The Impact of the Cruising Industry on Local Destinations.

Raosoft. (2004). Raosoft Sample Size Calculator. Raosoft, Inc., Seattle.

Soufriere Marine Management Authority – About Our Zones.

Spalburg, Jo. 2009. Cruise Market Study, VIP Consulting Group.

SLASPA (St. Lucia Air and Sea Port Authority). 2014. *Seaports: Cruise Schedules 2014/2015*.

St. Lucia National Report on the World Summit on Sustainable Development. 2001.

Sweeting, James, and Scott Wayne. 2003. *A Shifting Tide: Environmental Challenges and Cruise Industry Responses*. Interim Summary Report. The Center for Environmental Leadership in Business.

Symmonds, Mathew, William Hammitt, and Virgil Quisenberry. 2000. Managing Recreational Trail Environments for Mountain Bike User Preferences. *Environmental Management* 25 (5): 549–564. https://doi.org/10.1007/s00267 9910043.

Twining-Ward, Louise. 1999. Opinion Piece: Towards Sustainable Tourism Development: Observations from a Distance. *Tourism Management* 20: 187–188.

UNEP (United Nations Environment Programme). 2010. *National Environmental Summary St. Lucia*.

UNEP & WTO. (United Nations Environment Programme & World Tourism Organization). 2005. *Making Tourism More Sustainable: A Guide to Policy Makers.*

Weaver, David. 1993. Model of Urban Tourism for Small Caribbean Islands. *Geographical Review* 83(2): 134–140. Accessed January 14, 2021. https://doi.org/10.2307/215251.

Wilkinson, Paul. 2006. The Changing Geography of Cruise Tourism in the Caribbean. In *Cruise Ship Tourism,* edited by Ross Dowling, 170–181. CAB International.

Williams, Stephen. 1998. *Tourism Geography. Routledge Contemporary Human Geography Series*. London: Routledge.

CHAPTER 5

Tourism Resilience in the Caribbean Island of Cozumel: Best Practice and High-Risk Areas

Kennedy Obombo Magio

Introduction

The island of Cozumel, 30 miles long, is located off the eastern coast of Mexico's Yucatán Peninsula, opposite the coastal town of Playa del Carmen and near the Yucatán Channel. Before the COVID-19 pandemic, tourism and related commerce accounted for 78% of the island's economic activities (5 million visitors annually, Secretaría de Turismo—SEDETUR, 2020), with the remaining 22% derived from fishing, agriculture, ranching, manufacturing, and construction, Sistema Nacional de la Información Estadística del Sector Turismo de México—DATATUR, 2020). Cozumel is host to a mix of several types of visitors, including cruise ship "day trippers" who spend less than a day on the island; stay-over guests (some in all-inclusive resorts), mainly from the

K. O. Magio (✉)
CONACYT - Tecnológico Nacional de México/Instituto Tecnológico de Cancún, Cancún, Mexico, Quintana Roo
e-mail: kmagio@conacyt.mx

A. Lewis-Cameron et al. (eds.), *Managing Crises in Tourism*,
https://doi.org/10.1007/978-3-030-80238-7_5

United States, who typically stay three to four days; domestic travelers, low-budget backpackers (both national and international), and long-stay expatriates and retirees with homes on the island. Despite the strategic importance as Mexico's top cruise destination, policymakers, academics as well as conservation and development organizations are continuously questioning the resilience of tourism on the island and attempting to identify which strategies have been most successful for both conservation and local development goals. Empirical evidence from the present study reveals an area in need of attention; that of reducing negative environmental impacts—conciliating and linking environmental conservation and the potentially conflicting aims of tourism on the island. Cozumel´s fragile natural ecosystems include the Mesoamerican Barrier Reef, the Western Hemisphere's largest reef system, stretching from Isla Contoy at the tip of the Yucatán Peninsula down to Belize, Guatemala, and the Bay Islands of Honduras.

The overall objective of the present study is to identify both best practices and risk areas that are undermining Cozumel's resilience as a destination, with the main focus on environmental sustainability. Also, the researcher looks at key initiatives and projects to improve responsible tourism management practices in the island. The author seeks to provoke a rethinking of the current mode of developing tourism on the island and contribute to the body of knowledge on issues related to the scale and extent of tourism and the associated development costs and benefits; and to demonstrate that properly planned tourism can contribute to healthy livelihoods and ecosystems based on multi-sectoral partnerships. The assessment was conducted between December 2018 and August 2019 (a few months before the start of COVID-19 pandemic) using a carefully designed methodology that was based largely on stakeholder input and site visits, as well as an evaluation of the destination's policies and practices against the Global Sustainable Tourism Council—GSTC's Criteria for Destinations (GSTC-D). This included desk-based collection and analysis of destination policies and practices, focus groups, interviews, and a workshop with key destination stakeholders to assess the levels of implementation. Some 50 stakeholders from government, tourism businesses, NGOs, civil society, and academia were consulted during the study. The GSTC is the global standard-setting and accreditation body for sustainability of tourism businesses and destinations. GSTC came into being through efforts by the United Nations Foundation, together with the World Tourism Organization (UNWTO), the United Nations Environment Programme (UNEP), and the private sector.

To determine the degree of compliance with each of the GSTC-D indicators, the researcher analyzed the results according to four performance levels:

- Green = Documentation of policy exists, and it is being implemented in an exemplary manner—on a level with best practices;
- Yellow = Documentation of policy exists, and it is being implemented in an acceptable manner;
- Pink = Documentation of policy exists, but it is not being implemented—or vice versa and Red = No documentation of policy exists, and it is not being implemented. The high-risk areas according to the stakeholders 'priority include:

1. Building solid-waste management and reduction programs
2. Increasing safe water management and conservation
3. Developing a multi-year Strategic Plan and Vision for Sustainability
4. Creating a destination management organization (DMO)
5. Developing standards and training for tour operators and guides
6. Strengthening sustainability standards through certification, branding and marketing of sustainable tourism
7. Creating an energy management and conservation program
8. Expanding use of low-impact transport
9. Enhancing wastewater management programs
10. Developing an Integrated Monitoring System and public reporting of environmental, economic, social, cultural, human rights, and environmental issues related to tourism, and
11. Establishing programs to measure, monitor, minimize, publicly report, and mitigate greenhouse gas (GHG) emissions from tourism enterprises and infrastructure.

Tourism, Sustainability, and Resilience in the Wake of COVID-19 Pandemic: Insights from Mexico

As the global crisis surrounding COVID-19 plays out in real time around us, tourism is unquestionably in the spotlight more than ever. In Mexico, during the first three months of the COVID-19 pandemic, the industry came to a standstill, almost 2 million cruise passengers and revenues of 134.4 million dollars were lost, with the greatest impact in the Mexican Caribbean (SECTUR 2020). Experts estimate that it will take at least a couple of years for the sector to recover (Expansion 2020). In small islands like Cozumel which is dependent on cruise tourism, the COVID-19 crisis puts not only visitors at great risk but also the livelihood of local population.

In this recovery process, it has been recognized that many acute shifts and turns lie ahead in terms of how tourism is conceptualized, understood, performed, governed, and planned for, and topics related to sustainability and resilience of tourist destinations have never been so relevant. As well as presenting a range of new challenges which require informed and context-specific responses, the present moment is undoubtedly also an opportunity to build resilience, capacity, and positive transformations in destinations. This means that destinations should develop capacity for readiness, adaptability, and resilience to achieve first their own recovery, survival and prosperity, as well as, in turn, to nurture versatility in terms of how tourism activities can be employed to serve their host communities' needs in post-crisis recovery.

In this new transformative vision (one that puts priority on local community and the environment), sustainable tourism has been suggested by many scholars as the means for a steady and faster economic, environmental and social recovery (Băndoi et al. 2020; Everingham and Chassagne 2020; Galvany, Lew and Sotelo Perez 2020; OECD 2020; Lew 2018), recognizing that decades of unfettered growth in travel have put the world's treasured places at risk, particularly the environment. According to Galvany, Lew and Sotelo Perez (2020), and Băndoi et al. (2020), business activity is only one aspect of the "tourism system," as is the so-called tourism industry. At the very least, tourism also involves communities (e.g., residents), the natural environment and the lived environment (e.g., overtourism issues). The main focus in this recovery process is to make sure that tourism does not return to its former state where economic bottom lines and the pressure to meet the needs of the visitor override real aspirations and concerns for the local people (and for the natural environment for that matter). According to Galvany, Lew and Sotelo Perez (2020), tourism must consider inclusive social participation to democratize decisions and assume the responsibilities that the conservation of the natural and cultural heritage of a locality or region entails.

Methodology

A methodology similar to the one used by the World Wildlife Fund—WWF and the GSTC was used to implement the assessment in Cozumel using the GSTC Destination Criteria with focus on best practice and high-risk areas. The assessment covered four main themes: sustainable destination management, economic benefits to the host community, cultural benefits, and environmental benefits. The evaluation, scoring, and coding were done using a points and color method—the traffic light system—in which 1 and red indicate no documentation related to policies and no implementation; 2 and pink indicate documentation related to policies, but no implementation—or vice versa; 3 and yellow indicate documentation related to policies, and implementation; 4 and green indicate documentation on policies, implementation, and improving trends to the extent of best practice; and N/A is assigned when the indicator is not applicable in the destination. The methodology entails four main steps: 1. Desk-based assessment; 2. On-site assessment; 3. Reporting; and 4. Post-assessment implementation workshop.

Findings and Discussion

The four main themes for the criteria and indicators used in this study are noted in Fig. 5.1. Forty-one (41) criteria and the 113 indicators for meeting them are placed within these four categories. To determine the

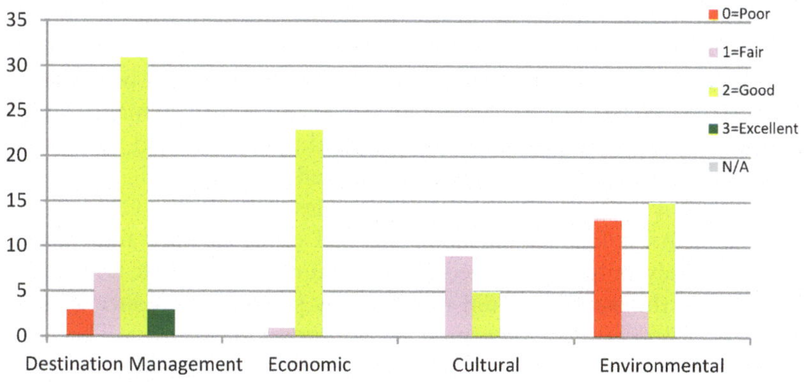

Fig. 5.1 Destination Sustainability Performance by Category

degree of compliance, the researcher evaluated and ranked each indicator according to four performance levels, designated by color: green, yellow, pink, and red. Of the 113 indicators analyzed, only three indicators (2.66%) were rated "green" for exemplifying best practices; a majority—74 (65.48%)—were rated "yellow," demonstrating that documented policies exist and are being implemented in an acceptable manner, with some room for improvement noted; 20 (17.70%) were rated "pink," signifying either inadequate policies or performance; and 16 (14.16%) were rated "red," signaling that no policies or practices exist.

High-Risk Topics and Recommendations

The researcher, based largely on stakeholder input and site visits, identified and made recommendations for the 12 most crucial high-risk sustainability and resilience issues. The top six risk topics as prioritized by Cozumel stakeholders are detailed below, with the related GSTC Criteria indicated in parentheses. These top six risk areas are ripe for stakeholders to develop action projects that would improve overall destination sustainability.

1. Solid-Waste Reduction (D10)

Cozumel does not have a well-functioning system to encourage enterprises and institutions to reduce, reuse, and recycle solid waste. In addition, not all of its residual solid wastes are disposed of safely and sustainably. Further, the waste collection system does not maintain public records of the amount of waste generated and how it is treated, although all waste collection trucks arriving at the sanitary landfill must be weighed. While Cozumel has not implemented a solid-waste management plan, it does have quantitative goals to minimize and ensure safe and sustainable disposal of waste that is not reused or recycled.

Although Centro de Acopio de Materiales Reciclables (CAMAR), a local municipal company, is making the best effort to store and transfer waste materials, only an estimated 7% of Cozumel's solid waste is recycled. In addition, the assessment did not find any program to reduce and/or eliminate plastic bottled water used by enterprises and visitors. There are facilities for waste disposal and recycling managed by Promotora Ambiental de La Laguna, S.A. de C.V. (PASA); however, their capacity is not adequate, and the sanitary landfill is relatively small in relation to the volume of solid waste generated in Cozumel. According to the participants, the concession given to PASA for waste management (sanitary landfill) services is expensive for the municipality but not as efficient as expected.

In some cases, the Municipal Government's Department of Public Services (La Dirección de Servicios Públicos Municipales) gives guidance to residents and tourists on minimizing and separating waste in public locations. In addition, there are guidelines for the collection, transport, treatment, and final disposal of solid waste in Cozumel (Reglamento para la Prestación del Servicio Público de Recolección, Transporte, Tratamiento y Disposición Final de Residuos Sólidos en el municipio de Cozumel). However, the assessment found that implementation of these regulations is problematic.

Recommendation: Build Solid-Waste Management and Reduction Programs

1. Develop an integrated system for waste management that will implement strategies for improving waste management and environmental quality in Cozumel. The system will enable the destination to optimize the available resources and attract funding for short-term and long-term projects in waste management.
2. The destination should develop an incentive program to reduce waste generation by all sectors and encourage enterprises and institutions to reduce, reuse, and recycle solid waste.
3. The destination should seek investments in infrastructure projects to guarantee 100% efficient waste management (solid waste, wastewater, and polluting gases), taking into account its volume and impact.

2. Water Management (D6), Water Security (D7), and Water Quality (D8)

Cozumel does not have a system to encourage enterprises to measure, monitor, reduce, and publicly report water usage. Also, there is no evidence that the impacts of tourism on local water sources are tracked, or that goals for reducing water consumption within the tourism sector are in place. The destination does not have a system to monitor its water resources to ensure that use by tourism enterprises is balanced with the water requirements of the host community. The present assessment did not find evidence of regular, reliable, and publicly available information on the water usage by either the host community or tourism service providers, such as hotels, whose operations require large quantities of water. The assessment found no evidence of efforts to measure and publicize information about the impact of tourism activities on the local aquifer or of efforts to reduce water usage by the tourism sector.

In addition, Cozumel does not have a management system to monitor drinking and recreational water quality using international quality standards. Although the Water Commission (Comisión de Agua Potable y Alcantarillado [CAPA]) conducts water quality studies at the source, the results of these monitoring exercises are not regularly available to the public. Consequently, the island does not have a system to respond in a timely manner to water quality issues. The most recent information about water quality and management on CAPA's website dates from 2016, and none of the government's published information, which starts from the third trimester of 2014, refers to the specific sustainability indicators outlined in the GSTC criteria.

Recommendation: Increase Safe Water Management and Conservation

1. Develop a water distribution and use plan to balance the present and future water consumption needs of Cozumel's residents and tourism enterprises. This should be based on a study to determine the water that can be available for tourism, without reducing the supply required by residents.
2. Conduct and publish a study on the quantity and quality of the island's aquifer, since there has been no diagnosis in the last 10 years. This should include techniques and costs for ensuring safe drinking water for the growing numbers of both residents and tourists in Cozumel and should draw lessons from successful initiatives in other similar destinations.
3. Develop strategic projects to improve the efficacy of aqueducts and water treatment plants and to replace old pipes in the municipality.
4. Create a program for water conservation in collaboration with international and federal organizations. This program should highlight the important financial savings to businesses that conserve water.

3. Sustainable Destination Strategy (A1)

Cozumel does not have an officially recognized multi-year destination strategy for sustainable tourism development. There are, however, short-term plans, strategies, and programs that partially incorporate aspects of sustainable tourism and address environmental, economic, social, cultural heritage, quality, health, safety, and protected-area issues.

Recommendation: Develop a Multi-Year Strategic Plan and Vision for Destination Sustainability

1. Develop an up-to-date Strategic Plan for Sustainable Tourism in Cozumel. Existing management plans, including the Sustainable Tourism Strategic Plan (PETSIC) 2014–2032, do not adequately address all components of sustainable destination management identified in the GSTC Criteria. Additionally, *PETSIC* was not formally adopted by the Municipal Government to guide destination management. It should be updated and integrated into this overall Sustainable Destination Strategy, which should include a vision for the development of the destination, and all key issues covered in the GSTC Criteria.
2. Additionally, the plan should be developed to take into account the Sustainable Development Goals (SDGs) and to integrate with other management plans at the state and national level. These include:
    - National Agreement for Tourism (*Acuerdo Nacional por el Turismo*)
    - National Development Plan
    - Sectorial Program for Tourism (Programa Sectorial de Turismo, 2019–2024).
    - The State Development Plan for Quintana Roo (*Plan Estatal de Desarrollo—Quintana Roo, 2016–2022*) and the Master Plan for Sustainable Tourism Development—Quintana Roo (*Agenda para el Turismo Sustentable—Quintana Roo*), which seek balanced development in state destinations.

4. Visitor Management (C2) and Visitor Behavior (C3)

The destination does not have a visitor management system for attraction sites that includes measures to preserve, protect, and enhance natural and cultural assets. However, a few initiatives are carried out by some tourism service providers and stakeholders—cruise lines, tour operators, and marine/underwater service providers—in relation to visitor management. In 2009, the Intersectorial Group (GI) launched a one-minute video in three languages (English, Spanish, and Maya) to create awareness among the tourists about the island's fragile ecosystem and to influence their behavior toward it. This video was widely circulated to local operators, such as hotels, restaurants, ferries, and tour companies. It is also important to note that visitor management actions by service providers are not uniform; some operators allow direct interaction with wildlife in sensitive sites. The evidence found in the assessment does not constitute an effective administrative mechanism that is responsible for implementing visitor management plans and operations, including wildlife and sensitive sites.

There are isolated cases where tour operators and service providers have codes of conduct to guide visitor behavior in all sites. In particular, for wildlife observation and interaction, their aim is to minimize and mitigate negative impacts in sensitive and fragile sites. Most of these best-practices companies have sustainable-management certification and regularly monitor compliance with their codes of conduct. The two ferry companies in Cozumel play short video clips on their vessels with messages about the importance of acting in a responsible way to protect the fragile island ecosystems. On the government side, the National Commission for Protected Areas (CONANP) has a code of conduct for the island's parks, however, there is no evidence of regular monitoring and compliance.

As a recommendation, the destination should have a visitor management system for attraction sites that includes measures to preserve, protect, and enhance natural and cultural assets. The system should include guidelines on physical carrying capacity. Before the COVID-19 pandemic, there were at least 3,000 cruise passengers on a normal day in Cozumel, walking or moving around the main town during the 5 to 8 h they were there, another significant number of travelers, including stay-over tourists, visited various attraction sites that are mostly sensitive and fragile. Therefore, setting an ecological carrying capacity for protected areas and coral reefs is required. Enforcement is also required in cases where the carrying capacity already exists. For example, Chankanaab National Park has a carrying capacity of 3,500 visitors per day and received between 800 and 1,000 during high season before the COVID-19 pandemic. However, these numbers would increase to 3,000 when cruise tourists are included. The development of a code of conduct for tour guides and operators and cultural and environmental guidelines for visitor behavior is recommended in all sites, and should include destination endorsement for wildlife viewing and interaction standards for both marine and terrestrial species. This information should be developed with community input, be consistently updated, and made publicly available.

5. Climate Change Adaptation (A5)

Cozumel does not have a current system for climate change adaptation and risk assessment. While there are laws and policies to mitigate climate change, their focus is not on the use of technologies to combat climate change. This indicator was, therefore, rated "pink." It is recommended that the local government updates its climate change adaptation policies and strictly enforces design and building regulations covering building height limits, installation of alternative sources of energy for generators and other important equipment, safe storage of harmful or hazardous substances, and appropriate design and siting of waste treatment plants. The assessment determined that Cozumel also does not have a specific program to educate and raise awareness among the public, tourism enterprises, and visitors about climate change. This information is, however, partially included in the Municipal Communication Plan that seeks to engage the local community on a range of issues affecting the island. Discussions with some stakeholders, including Cozumel residents and visitors, revealed that they are not sufficiently aware of the potential impacts that climate change could have on the island, including on its tourism sector.

The present study recommends the creation of a Climate Change Management Unit (CCMU) or Coastal Zone Management Unit (CZMU) within the DMO, bringing together experts, planners, and adequate budgetary allocation from the local government or outside. As a model, Cuba has succeeded in integrating coastal tourism management with climate change mitigation strategies. Cuba has teams of technicians, scientists, and environmentalists dedicated to management and monitoring of the country's 3,000 miles of coastline. It has also committed resources to create the appropriate legal, institutional, and political frameworks to forecast and mitigate impacts associated with climate change.

Findings of Good Practices

This section includes indicators rated "green," i.e., documentation of policy exists and implementation is exemplary, on a level with international best practices and "yellow," indicating that documentation of policy exists, and implementation is acceptable. The observations below are based on information collected during the assessment (both desk-based and on-site, including stakeholder meetings and site visits) and are categorized by criteria that can contain multiple indicators.

1. Tourism Seasonality Management (A4)

Cozumel has exemplary measures for mitigating the seasonal variability of tourism. These include efforts to balance the needs of the local economy, community, cultures, and environment and to identify year-round tourism opportunities. The results of the Rapid Assessment in 2012 brought to the fore many opportunities for tour operators and investors. Tapping into that, the Riviera Maya Promotion Trust (Fideicomiso de Promoción Turística de la Riviera Maya) works on diversifying tourism offerings and marketing off-season events year-round. For instance, Cozumel has experienced growth in the number of cultural tourists and bird watchers. The German Agency for International Cooperation—GIZ is also supporting an initiative called Businesses for European Tourism (Empresarios Por el Turismo Europeo, or EPTE), which seeks to increase the number of European tourists visiting Cozumel. It does so by working on specific components of the tourism offer that attract, for example, cultural tourism. In the 2012 Rapid Assessment, cultural tourism as part of the product was rated red. Although these efforts do not form part of a specific marketing strategy envisaged in the GSTC criteria, they have been significant in mitigating seasonal variability in tourism. All Promotion Trusts in Quintana Roo were dissolved by the end of 2018 to form the new Tourism Promotion Council, Quintana Roo, in which all municipalities have a permanent seat. It will be necessary to approach this council with a long-term marketing strategy for Cozumel that is guided by the principles of sustainability and based on the unique natural and cultural heritage characteristics of the destination—and its recent designation as a biosphere reserve by UNESCO.

2. Planning Regulations (A7)

Good practice was found in relation to planning and regulation (rated "yellow"). The destination has planning guidelines, regulations and/or policies that require environmental, economic, and social impact assessment and that integrate sustainable land use, design, construction, and demolition. Some of these guidelines, regulations and policies designed to protect natural and cultural resources are local, while others are state and federal. The main challenge facing these regulations and policies is enforcement. Most impressive is the existence of five protected areas (three marine, one terrestrial, and one marine/terrestrial), with each one having its own management plan (plan de manejo). Also, Cozumel has an Ecological Zoning Plan (Programa de Ordenamiento Ecológico Local del Municipio de Cozumel—POEL). Recently, Cozumel was designated as a Biosphere Reserve by UNESCO under the Man and Biosphere Program. This recognition comes with adequate planning to ensure that development and conservation objectives are integrated. There is a need to harmonize the existing Ecological Zoning Plans (Urban Development Plan—PDU, Risk Atlas, Laws on Human Settlement, POEL) with Management Plans for the Protected Areas. Sustainable land, marine, coastal, and freshwater use, design, construction, and demolition, are guided by the Ecological Zoning Plan for Cozumel (POEL) and the Construction Regulation (Reglamento de Construcciones para El Municipio de Cozumel del Estado de Quintana Roo).

Training programs do not yet exist to teach communities about the sustainable use and management of their natural resources for commercial purposes. However, there are initiatives to promote local handicrafts based on sustainable use of natural resources. Awareness campaigns have been carried out over the last four years to promote, for example, the eradication of lionfish, reduction in the use of drinking straws, and reducing the use of plastic bags.

3. Sustainability Standards (A11)

Cozumel has industry-supported sustainable tourism certification or environmental management systems; some tourism service providers—particularly hotels and diving shops—have sustainability certifications and environmental management systems, but not all. Some of those sustainable tourism certifications or environmental management systems are recognized by the GSTC (Control Union, EarthCheck, Rainforest Alliance and Ecotourism Australia). About 17 beaches have Clean Beach Program Certification; 19 establishments have health and hygiene certification for food and beverage establishments (Distintivo H); 16 restaurants, 1 hotel, 9 travel agencies, and 24 other tourism establishments have service quality certification (Distintivo M); and 31 local tour guides have been certified to *NOM-09-TUR-2002* by the federal Ministry of Tourism (SECTUR). While there is a need to monitor tourism business participation in tourism certification or environmental management systems, there is no publicly available list of sustainability certified or verified entities.

4. Safety and Security (A12)

The assessment found evidence of ongoing compulsory inspections of fire, food hygiene, and electricity safety for tourism properties; there are health inspections by COFEPRIS and the Ministry of Health (SESA). The Civil Protection Department (Protección Civil) carries out mandatory electric security inspections in all tourism properties, depending on the area of service provision; they check lighting, gas pipes and cylinders, fuel, and so on. Some stakeholders expressed concern over the growing numbers of security incidents, particularly house robberies; however, the Municipal Government has a system in place to prevent and respond to crime. There is public reporting of safety and security: the Municipal Government security department has a Facebook account where security updates are made and members of the public can comment and report incidents. The director of security has radio sessions every Friday to discuss the security situation on the island and answer any security-related questions from the public. A taxi licensing system known as "concesión" has clear pricing and an organized taxi dispatch system at points of tourist entry, but the prices are not always respected.

5. Crisis and Emergency Management (A13)

Cozumel has crisis and emergency response plans that consider the tourism sector and are publicly available. For island destinations, crisis management aimed at hydrometeorological phenomena is crucial, and Cozumel's crisis and emergency response plans for hurricanes have been praised by experts in the field as some of the best in the entire country. However, similar plans should be made for all likely threats to the destination, including health crises. There is financial and human capital to implement the crisis and emergency response plan, but it is minimal. There is evidence that the crisis and emergency response plans for hurricanes were developed with input from the tourism private sector and include communication procedures for during and after such a crisis or emergency.

6. Environmental Risks (D1)

Cozumel has successfully identified the majority of environmental risks in both terrestrial and marine sites through sustainability assessments carried out by third parties over the last few years. This is a major improvement from 2012, when the Rapid Assessment was carried out by STI; at that point, stakeholders though that not all risks had been sufficiently identified and evaluated. However, there is no system in place to address identified risks. It is also important to recognize the amount of work done on the subject in the last 20 years by the academic community: for example, the Risk Atlas for Cozumel recently published in a scientific journal ("Atlas de Riesgos de la Isla de Cozumel, México," *Teoría y Praxis,* Núm Especial, 74–93) and the Risk Atlas prepared by local biologist Nicolás Carrillo Fajardo (*Atlas de Riesgos del Municipio de Cozumel,* 2011). The public sector, as well as local non-governmental and civil organizations, and national and international organizations such as GIZ have done an equally great job identifying and managing environmental risks in Cozumel. There is a growing social awareness and interest and commitment among key stakeholders; many recognize this criterion as one of the most important for Cozumel, because the main attractions of the island are totally dependent on the good health of the fragile ecosystem. During the workshops, participants expressed concerns over risks that are caused by contaminating wastes (solid, liquid, and gas) on the underground, aquifer, and reefs. These risks have not been adequately identified and evaluated to know their frequency and impact size. In addition, the destination does not have an integrated system in place to address them. Equally, participants expressed concern over the possible impacts of secret dump sites, untreated water, used oil and emissions of contaminant gases, whose effects have been partially measured by the academic community.

7. Protection of Sensitive Environments (D2)

The destination has largely succeeded in various aspects of conservation, in part because almost half of the island is covered by five legally recognized protected areas:

1. Cozumel Reef National Park—11,988 hectares, federal protection (Zona Arrecifes de Cozumel—Área Natural Protegida/Programa de Manejo)
2. Chankanaab Lagoon National Park—14 hectares, state protection (Laguna de Chankanaab, Cozumel—Parque Natural)
3. Colombia Lagoon—1,131 hectares, state protection (Laguna Colombia—Área Natural Protegida, con la categoría de Zona Sujeta a Conservación Ecológica, Refugio Estatal de Flora y Fauna)
4. Jungles and wetlands of Cozumel (Selvas y humedales de Cozumel)—37,829 hectares, federal protection
5. Flora and Fauna Protection Area for Northern Cozumel (Área de Protección de Flora y Fauna—Norte de la Isla de Cozumel)

In 2016, Cozumel was designated as a Biosphere Reserve, joining the prestigious list of areas recognized by UNESCO under the Man and Biosphere Program (MAB) for balancing conservation and development objectives. Despite this significant progress, protection of sensitive ecosystems is not totally guaranteed, owing to the growing pressure from tourism development. The urban sprawl in Cozumel is estimated to be 38 hectares per year, mainly caused by the migration of people attracted by job prospects in tourism. The destination has maintained an inventory of sensitive, threatened, and protected status of wildlife, marine life and habitats; in the case of Cozumel's marine and terrestrial protected areas, the inventory is a complete one. Overall, there is a need to integrate and regularly update similar information generated by different organizations: public and private sector, the academic community, NGOs and civil organizations.

The assessment did not find a management system to monitor impacts and to protect ecosystems, sensitive environments, and species. However, there are significant efforts by CONANP, NGOs and civil organizations who work together with WWF-México, local cooperatives, the community, the Municipal Government, and some tourism service providers. The main challenges in this aspect include limited financial resources to monitor the well-being of ecosystems, the absence and sometimes inefficiency of government institutions whose responsibility is to enforce conservation and protection laws (PROFEPA), and the lack of civic awareness of the local community to monitor, report, and demand punishment for those who break the law.

The destination does not have an integrated system that prevents the introduction of invasive species, but there are successful initiatives and programs aimed at combating invasive species. For example, efforts to eradicate the invasive lionfish has been largely successful; but efforts to eradicate the invasive species of the Australian pine tree (*Casuarina equisetifolia*) have not enjoyed similar success rates.

Conclusion

The present assessment generated mixed results, issues that are undermining the destination's environmental, social, and economic sustainability and recommended improvements for responsible tourism practices on the island have been identified. This work builds on earlier studies, including the 2012 *Rapid Sustainable Cruise Destination Diagnostic* (2012 Rapid Assessment) which was part of ongoing initiatives in Cozumel to demonstrate that properly planned tourism can contribute to healthy livelihoods and ecosystems based on multi-sectoral partnerships. As seen in the results, the environmental category had the highest number (13) of indicators that scored red (poor), areas found to be at risk include: wastewater management (D9), energy conservation (D5), solid-waste management (D10), low-impact transportation (D12), water quality (D8), water security (D7), water management (D6), greenhouse gas emissions (D4), and light and noise pollution (D11).

A majority of the indicators across all categories (74) were rated yellow (good). While this demonstrates positive progress on sustainability issues, it also signals that there is more work to be done to move these indicators into the green, or compliance with best practices, category. A similar result was found in the 2012, Rapid Assessment. In fact, some indicators that were favorably rated in 2012 received a lower rating in the present assessment. This included wastewater management (D9). The variation can be explained by increasing resource pressures as Cozumel's resident population grew from 79,500 in 2010 to about 86,400 in 2015, it was estimated to be well over 100,000 in 2020. In addition, urban sprawl in Cozumel is estimated to increase 38 hectares per year due to the migration of people attracted to Cozumel by job prospects in tourism. Also, the number of tourism enterprises and tourists has grown tremendously since the last assessment in 2012.

It is expected that the assessment results, which include important recommendations, will assist Cozumel in addressing the identified risk areas and implementing key initiatives and projects through strategic partnerships and prioritized investments. The outcomes should be used to inform and shape policy and planning in other Mexican destinations as well as those throughout the Mesoamerican Barrier Reef. The recommended Strategic Plan, as a key instrument of governance on which other reforms can be built, should be designed to transcend political parties and to carry over from one government administration to the next in Cozumel. Broader and deeper collective action remains critical to achieving the three primary destination stewardship goals: (1) provide sustainable livelihoods; (2) conserve the resources for future generations; and (3) create a high-quality visitor experience.

References

Băndoi, A., E. Jianu, M. Enescu, G. Axingte, S. Tudor, and D. Firoiu. 2020. *The Relationship between Development of Tourism, Quality of Life and Sustainable Performance in EU Countries, Sustainability*, 12(1628) MDPI, pp. 1–24, https://doi.org/10.3390/su12041628.

Everingham, P., and N. Chassagne. 2020. Post COVID-19 Ecological and Social Reset: Moving Away from Capitalist Growth Models Towards Tourism as Buen Vivir. *Tourism Geographies* 22 (3): 555–566, Routledge Taylor & Francis Group. https://doi.org/10.1080/14616688.2020.1762119.

Expansión (07.08.2020). El COVID-19 le ha costado a los cruceros 2 millones de pasajeros y 134 mdd. Sección Empresas. Retrieved from https://expansion.mx/empresas/2020/08/07/el-covid-19-le-ha-costado-cruceros-2-millones-de-pasajeros.

Galvany, A., A.A. Lew, and M. Sotelo Perez. 2020. COVID-19 is Expanding Global Consciousness and the Sustainability of Travel and Tourism. *Tourism Geographies* 22 (3): 567–576. https://doi.org/10.1080/14616688.2020.1760924.

Lew, A.A. 2018. Why Travel? – Travel, Tourism and Global Consciousness. *Tourism Geographies* 20 (4): 742–749. https://doi.org/10.1080/14616688.2018.1490343.

Martínez, O. F., T. Ihl, and J. Rojas López. 2016. 'Risk Atlas for Cozumel Island (Atlas de Riesgos de la Isla de Cozumel, México)'. *Teoría y Praxis.* Núm. Especial, pp. 74–93.

Nicolás Carrillo Fajardo, N.C. 2011. *Risk Atlas for Cozumel (Atlas de Riesgos del Municipio de Cozumel).* Licitación No. LO-823001999-N3–2011. Numero de obra 123001PP03811. Número de expediente P11/23001/AE/1/031. Municipio de Cozumel, Estado de Quintana Roo & Secretaría de Desarrollo Social, Gobierno Federal.

OECD – Organisation for Economic Conservation Development. 2020. *Rethinking tourism success for sustainable growth, Trends and Policies*, Chapter 3. Retrieved from https://www.oecd-ilibrary.org/docserver/82b46508-en.pdf?expires=1599912610&id=id&accname=guest&checksum=6E3F44FE682E9F1D4417856A8AB17D56.

Secretaría de Turismo- SEDETUR. 2020. Indicadores Turísticos. Available at http://qroo.gob.mx/sedetur/indicadores-turisticos. Accessed January 7, 2020.

SECTUR – Secretaría de Turismo. 2008. Agenda 21 para el Turismo Mexicano. Un marco de acción para el desarrollo sustentable de la actividad turística. Retrieved from https://moodle2.unid.edu.mx/dts_cursos_mdl/lic/AET/TC/AM/02/Agenda21.pdf.

SECTUR – Secretaría de Turismo. 2020. Programa Sectorial de Turismo 2020–2024. Retrieved from http://www.sectur.gob.mx/gobmx/wp-content/uploads/2020/07/PROSECTUR_2020-2024.pdf.

Sistema Nacional de la Información Estadística del Sector Turismo de México – DATATUR. 2020. Actividades en Crucero. Available at: http://www.datatur.sectur.gob.mx/SitePages/Actividades en Crucero.aspx. Accessed May 7, 2020.

CHAPTER 6

A Resilient Eco-Tourism Island: A Case Study of Dominica and Its Tourism Recovery Strategies Post 2017 Hurricane Maria

Tenisha Brown-Williams and Amanda Charles

Introduction

Resilience continues to be a primary discussion in relation to Small Island Developing States (SIDS) based on their inherent vulnerabilities and susceptibility to external shocks. For Caribbean SIDS these shocks often manifest themselves in natural disasters such as hurricanes, flooding, earthquakes and volcanoes resulting in millions of dollars in infrastructural damage, destruction to critical vegetation, wildlife and coastlines, unfortunate loss of life and overall economic decline. Ötker and Srinivasan (2018, 40) state, "average estimated disaster damage as a ratio to

T. Brown-Williams (✉)
University of the West Indies, St. Augustine, Trinidad and Tobago
e-mail: Tenisha.brown@sta.uwi.edu

A. Charles
Sustainable Tourism Division, Caribbean Tourism Organization (CTO), Bridgetown, Barbados
e-mail: acharles@caribtourism.com

A. Lewis-Cameron et al. (eds.), *Managing Crises in Tourism*,
https://doi.org/10.1007/978-3-030-80238-7_6

GDP was 4.5 times greater for small states…but six times higher for countries in the Caribbean." They also note that the Caribbean is "…seven times more likely to be hit by natural disasters than larger states and twice as likely as other small states" (p. 40). As a result, the Caribbean has incurred losses in excess of US$136 billion dollars over the last two decades due to natural disasters (Association of Caribbean States 2017) with many affected territories taking years to rebound and return to economic stability. Further compounding the situation is an overall dependence by the region on tourism for economic growth and development. By nature, tourism is a viable yet volatile economic activity and is also extremely sensitive to external shocks. At the time of this writing, the Caribbean is grappling with the devastation of regional tourism due to the advent of the novel coronavirus (COVID-19) pandemic and preparing for an earlier than normal start of the hurricane season in May 2021. Therefore, there has been heightened discussions surrounding best practices on disaster recovery and the need for Caribbean SIDS to build resilience during these times of extreme uncertainty. Following the devastating 2017 Hurricane Season, the island of Dominica announced its intention to be the "world's first climate-resilient nation" (Lieberman 2020), the hurricanes having an almost debilitating impact on infrastructure, agriculture and the tourism industry key pillars of the island economy and indicators of socio-economic growth. With this in mind, this chapter will examine the tourism recovery efforts of Dominica, following the passage of Hurricane Maria in 2017, which made landfall as a dangerous Category 5 Hurricane, with the intent of extracting key lessons that can potentially be applied to the destination and Caribbean region at large, in post COVID-19 tourism recovery actions. The chapter will firstly provide a brief background of the tourism sector and the effects of Hurricane Maria in Dominica and will then review main discussions in literature on tourism recovery and resilience. Findings from structured interviews will be presented and discussed with key lessons and recommendations forthcoming.

Literature Review

Natural Hazards and SIDS

SIDS are no strangers to natural hazards with some incurring annual damages of US$0.5–US$1 billion per year, resulting in major disasters

(Bettencourt et al. 2014). These disasters occur in the form of earthquakes, hurricanes, volcanic activity, tsunamis, torrential rain, flooding and storm surge. For SIDS, their location in vast oceans compounded by their small land mass renders them exposed and vulnerable to natural weather patterns and marine processes. Additionally, their formation based on volcanic activity and other geological phenomena predisposes them to seismic activity. Climate change issues such as rising sea levels, intensification of storms and other unusual weather patterns now exacerbate the inherent vulnerabilities and threaten the very existence of small islands states, earning them the title of "the most disaster-prone countries in the world" (Slany 2020). While the odds based on geospatial composition seem to be naturally against SIDS, regrettably it is further amplified by their socio-economic characteristics. Small island states are also challenged by a narrow resource base resulting in lower economies of scale (OHRLLS 2011); heavy reliance on public sector services with large public sector spending (OHRLLS 2011; OCED 2018); limited opportunities for private sector growth (UN-OHRLLS 2011); high dependence on foreign aid/intervention (Everest-Phillips 2014) as well as high levels of debt and lack of human capital (UNEP 2014). These limiting factors pose as barriers to fully equip SIDS to prepare for, recover from and withstand future impacts of disasters on their economies and population.

Among the many impacts of climate variability and climate change (CVC) is an increase in the intensity and severity of natural hazards (Simpson et al. 2008). The realization that natural hazards are not expected to diminish but rather worsen is a sobering one for Caribbean SIDS in particular, and for the health of their tourism dependent economies. Simpson et al. (2008) indicated that Caribbean CVC could result in a wide range of detrimental consequences to the tourism industry, including:

(a) An increase in the frequency and intensity of extreme weather events (droughts, floods and hurricanes), which would result in damage or loss of infrastructure, increased insurance costs or even lost insurability, business disruption and evacuation costs, as well as a negative image of the Region as a safe destination;
(b) Sea level rises accelerating coastal erosion and destabilizing infrastructure which would increase the vulnerability of tourism facilities in coastal areas (beaches, yachting marinas and cruise ship piers, a large percentage of accommodations, heritage attractions);

(c) Saltwater intrusion to coastal aquifers would threaten freshwater supplies, with the Bahamas being identified as the world's most vulnerable nation to sea level rise by the percentage of land area lost (Dasgupta et al. 2007);
(d) Temperature changes, resulting in warmer winters in northern markets and warmer summers in the Region, affecting seasonal demand;
(e) Changing precipitation patterns, leading to reduced water supply; and
(f) Increased sea surface temperatures causing coral reef bleaching and mortality (IUCN 2008).

Bueno et al. (2008, 3) predict that with "…increased hurricane damages, loss of tourism revenue, and infrastructure damages—the Caribbean's annual cost of inaction is projected to total $22 billion annually by 2050 and $46 billion by 2100." These costs represent a range of 10–22 per cent loss to the Caribbean economy. They further state "as temperatures rise and storms become more severe, tourism—the life-blood of many Caribbean economies—will shrink and with it both private incomes and the public tax revenues that support education, social services and infrastructure (Bueno et al. 2008, 2). The impacts of natural disasters further increase migration of locals into mainland countries and the "brain drain" phenomena that they incur on a cyclical basis (Wetzel et al. 2012). Moreover, the tourism products that are primarily developed around natural resources are compromised which according to Thomas and Lindo (2019) is problematic since coastal-based tourism contributes more than 50% of the GDP for tourism dependent islands.

The outlook for SIDS is not all grim with many researchers counter-arguing that their inherent vulnerabilities foster competitive advantages (Briguglio 1995; Easterly and Kraay 2000; Armstrong and Read 2002; Croes 2006; Baldacchino and Bertram 2009; Philpot, Gray and Stead 2015). Furthermore, it is noteworthy that there is variation in the level of vulnerability to natural hazards and disasters among SIDS with some handling disasters better than others (ECLAC 2011; Sjöstedt and Pivotkina 2017). Thomas et al. (2020, 6) aptly states that it is "…very much dependent on context and scale." Moreover, researchers continue to do the needful to find plausible solutions since geographical re-location is unfathomable. Solutions such as adaptation, mitigation and disaster risk reduction (DRR)/disaster risk management are among some measures proposed to minimize the devastating impacts of natural disasters (Betzold 2015; Pacific Community 2016; CTO 2020). These solutions are however contingent on access to financial, human and knowledge-based resources and the coherent management of all stakeholders to achieve the best possible outcomes.

The Caribbean Tourism Organization (2020) in the development of a recent publication titled "Multi-Hazard Risk Management for the Caribbean Tourism Sector" identified the disaster risk management cycle (DRMC) including the need for tourism destinations to develop appropriate recovery and response measures post disaster to enhance future preparedness. The UNCTAD (2021) states that the main aim of these solutions should be to build resilience referring to it as the best antidote to address and manage environmental and economic challenges. The subsequent discussions will explore literary perspectives on disaster response and recovery measures in the context of tourism and key considerations for building a resilient tourism industry for Caribbean? SIDS.

Tourism Recovery

For tourism-dependent countries the ability to restore and rebuild their tourism sector in a short-time frame is critical to their economic survival. Recovery in the tourism sector is linked to a return to normalcy measured by an enhanced feeling of safety, repair of infrastructure and a return to pre-disaster arrival figures, which occurs over multiple timeframes (Khazai et al. 2018). Tourism recovery is not viewed as a onetime occurrence but rather a process that occurs over the short–medium–long term period or on a phased basis (Faulkner 2001; World Bank 2020; Reddy et al. 2020). Reddy et al. (2020) in their development of a conceptual framework for post conflict tourism destination development identifies the Phoenix and Normalization phases. The Phoenix phase is described as the interim position a destination may take to adapt to the effects of a long-term crisis. It involves re-imaging, transforming and innovating of tourism assets the destination was well known for pre-disaster. Conversely, the Normalization phase avoids the Phoenix phase with the intent of transitioning speedily and directly from the disaster to the normal environment that existed pre-disaster. This transition is predicated on pre-disaster/proactive policies and plans that can be easily activated such as the development of "safe" tourism zones and a niche product portfolio with supporting marketing strategies.

The process of recovering from a hazard or disaster is contingent on the type, severity, scope, origin and duration of the impact (Ritchie and Jiang 2019). This can be observed in instances where earthquakes occurring at varying magnitudes cause less or more damage based on the depth of the earthquake or where hurricanes depending on their intensity and speed cause minimal or extreme devastation. Indeed, the more severe the impact, the slower the rate of recovery can be for the location. At the time of this writing, the La Soufriere Volcano in Saint Vincent and the Grenadines began to erupt in December 2020 with an explosive eruption occurring on April 9, 2021. The extended crisis has had an adverse impact on the economy of the island state with predictions that losses can amount to 50 per cent or more of GDP for 2021 if the situation prolongs (Wyss 2021). While the rate of recovery is influenced by disaster typologies and other external factors, there are internal factors that influence how quickly a tourist destination recovers from the impacts of a natural hazard/disaster. Many researchers are of the view that the rate of recovery is dependent on planning and preparatory measures taken by the destination beforehand (Faulkner 2001; Hystad and Keller 2008; Ritchie 2008; Mair et al. 2016; Gurtner 2016). These measures include disaster management forecasting and planning covering items such as risk assessments, development of protocols, communication strategies, warning systems, creation of disaster management teams, identification of command centers and contingency plans. Also essential in pre-planning actions is the identification of key stakeholders and their relevant roles and responsibilities. Hystad and Keller (2008, 161) state:

> Tourism disaster management should incorporate the mandate and potential of the stakeholders involved throughout the disaster planning, response, and recovery periods. Cooperation between stakeholders is a key factor of disaster management; therefore, it is important to clarify responsibilities during all stages of a disaster.

Beirman (2018) credits Thailand's remarkable tourism growth even after extreme natural disasters in 2004/2005 and 2011 to the existence of a ready-made strategy available for deployment when the disaster strikes. He closely aligns their approach to sentiments shared by Scarpino and

Gretzel (2015) which focuses on proactive planning, strategic development, evaluation and control and proactive risk management. These among other factors he determined accounted for Thailand's ranking among the top ten most visited countries in the world in 2017, even though the effects of earlier natural disasters were further compounded by civil upraising and terrorist attacks in 2014 and 2015 respectively. Thus, it is clear according to Ritchie (2008) that an "...understanding of tourism disaster planning, reduction and readiness and the development of suitable policies and initiatives from tourism agencies and industry associations to facilitate more effective planning" is critical to tourism recovery.

Destinations may have effective disaster planning processes and still not respond effectively when the disaster occurs. Additionally, it is possible that pre-planning activities become partially or wholly irrelevant depending on the nature of the disaster event. Therefore, real time response to the disaster whether based on existing or amended pre-planned measures must be effective. In many instances, recovery plans are formulated based on the nature of the disaster pointing to the need to be adaptable and flexible. Fakhruddin et al. (2020) identify effective responses during a crisis as inclusive of efficient and effective information dissemination, transparent governance and collaborative structures. This includes having modern information technologies and well-developed communication channels, dissemination of information to targeted population in a transparent manner, strong community vigilance and collaboration, coupled with evidence-based decision-making and continued support. They identified New Zealand as a destination exhibiting effective responses during COVID-19 and compared the response of Dominica after the hurricane struck deeming it to be ineffective, due to poor technology and fragmented communication channels, weak community vigilance, lack of public education measures, top-down governance and bureaucratic structures, and inadequate and inconsistent information or misinformation.

Likewise, the implementation of crisis communications strategies as part of the destination's response measures are also crucial for rapid recovery. This features in Ritchie's (2004) Tourism Crisis and Disaster Management Framework which emphasizes the control of crisis communication and the dissemination of accurate information in a timely manner to combat falsehoods. This involves among others, having a communication strategy and appointing a spokesperson. Zhang et al. (2021)

determine that this is important for managing the destination's image, brand or re-branding process, which is a critical factor for domestic and short-haul market recovery. These sentiments are exemplified in the case of Israel whose response and recovery from an extended crisis between 2003 and 2006 was applauded partially due to a coordinated strategic marketing approach, centralized, accurate and credible information service, and implementation of a media and public relations marketing campaign to challenge negative perceptions with positive facts and testimonials (Beirman 2009). Regarding Thailand, Beirman (2018, 284) also highlighted that "...TAT's (Tourism Authority of Thailand) crisis communication contributed to rapid recovery."

Tourism Resilience

Resilience is regarded as the ability of an entity (society, community, region) to cope, absorb and withstand external and internal shocks (Adger 2000; Rose 2004; Haimes 2009; Simmie and Martin 2010; Hosseini et al. 2016). For tourism dependent economies, the nature and extent of resilience is connected to their ability to adapt to internal and external changes, immediate and otherwise (Luthe and Wyss 2014). An ECLAC (2011, 27) report which sought to evaluate resilience of Caribbean SIDS under three dimensions (economic, environmental and social) discovered that across all dimensions, resilience differed based on several conditions including the degree of exposure to a particular shock and concludes that "....no single mechanism for building resilience to these impacts is applicable across the entire subregion."

Cheer and Lew (2018) point to the scale and rate of change as key factors in resilience but also recognize that resilience is contingent on the implementation of internal practices inherent in issues of management, resources, planning and governance. Their conceptual framework focuses on pre-exiting measures that build resilience, such as the existence of social and economic support systems, economic training, diversification and innovation, proper management programs and strategies targeted at the conservation of nature, culture and human welfare, all of which are dependent on cohesive relationships that are formed between government, industry and community. This is supported by Sharma et al. (2021, 8) who proposes as part of their resilience-based framework that "...three segments, namely, governments, market players and local communities, need to get their act together to lend resilience to the industry." They also identify the need to strengthen the use of technology to promote flexible and adaptable thinking. The work of Cochrane (2010) expands on the Sphere of Tourism Resilience framework which includes three (3) principal considerations inclusive of leadership, stakeholder cohesion and the harnessing of market forces. Leadership calls for the destination to have a clear vision and to "...create structures to resolve conflicts over resource use, drive forward change and spark cohesiveness and market engagement among stakeholders" (Cochrane 2010, 10). Stakeholder cohesion calls for the effective coordination of private, public and third sectors while the model also identifies the need to understand and engage market forces effectively. Broader elements within the model includes flexibility, learning and adaptability as key features which combine to undergird the three (3) principles.

Beirman (2018) in his application of the framework in a study on Thailand's approach to destination resilience, was able to determine that two (2) main factors influenced their reputation as the world's most resilient tourism destination, that of, speed of response in communicating with visitors and stakeholders, and the ability to bring a diverse group of stakeholders together to collaborate and act for the common good. According to Beirman (2018, 291), "Thailand's strength as a resilient tourism destination has been TAT's speed and professionalism in responding to the challenges when they arise and frequently having a ready-made strategy to employ." Beirman noted that these actions encompass anticipatory measures, short-term coping responses, long-term adaption and responsiveness, which are core characteristics of a resilient tourism destination that is continually adapting to changes in the internal and external environments.

Dominica: An overview

The Commonwealth of Dominica is a 750 square kilometer island located in the eastern Caribbean Sea between the French territories of Guadeloupe and Martinique. As of 2018, a population size of 72,000 was recorded (Commonwealth Secretariat 2021) consisting of indigenous Carib Indians and citizens with European and African ancestry (Discover Dominica Authority 2021). Owing to its volcanic origin, the island enjoys unique geological features such as the Boiling Lake, the second largest active flooded fumarole in the world. Dominica has gained renown for its mountainous terrain abounding with rivers and waterfalls, natural hot springs and tropical rainforests. The island is also home to many rare endemic plants, animals and bird species. These characteristics have been used to position Dominica as the "Nature Island of the Caribbean" and forms a critical part of its tourism sector.

Following independence from British rule, Dominica, like many other islands vigorously pursued agriculture including the exportation of bananas as a means of generating revenue. However, economic growth premised on this was stymied due to a decline in demand in the European market (Caribbean Community (CARICOM) Secretariat 2016) and the ultimate collapse of the banana trade (Holladay and Powell 2016). Today, agriculture still contributes significantly to the economy of Dominica through the production of alternative crops for exportation most of which are used for inter-regional trade however, much focus has been placed on leveraging the island's natural resource for tourism purposes. In 2019, international tourism revenue as a percentage of Dominica's Gross Domestic Product (GDP) was 19 per cent while 64 per cent of total exports was from international tourism (UNWTO 2020). A 2020 report by the Inter-American Development Bank (IDB) ranked Dominica among the top ten most tourism dependent countries in the world through an analysis of total export receipts, output as a share of real GDP, and employment as a share of total national employment for the period 2014–2018 (Mooney and Zegarra 2020). Arrivals to the destination from 2014 to 2019 averaged around 77, 000 visitors per year in stayovers and 228,000 per year in cruise stops (Invest Dominica Authority 2021). Average yearly revenue generated from stayover and cruise arrivals is estimated at US$107 million between 2014 and 2018 (Invest Dominica Authority 2021) and the total contribution to employment by tourism for the same period was 33 per cent (Mooney and Zegarra 2020).

The Impact of Hurricane Maria

In mere hours, Hurricane Maria which had been churning in the Atlantic Ocean for thirty hours prior to its landfall in Dominica, moved with rapid intensification from a category 1 to a category 5 hurricane (National Weather Service 2017). On September 19, 2017 at approximately 9:15 pm Hurricane Maria struck the island leaving in its wake unprecedented destruction amounting to EC$2.51 billion (US$931 million) in total damages and losses of EC$1.03 billion (US$382 million)

(Government of the Commonwealth of Dominica 2017), and an unfortunate death toll of 31 dead and 37 missing. According to the World Bank (2017) total damages and loss amounted to an incomprehensible 224% of GDP. Moreover, the tourism sector suffered some of the greatest loses after the housing sector and "the rainforest was destroyed, having a severe impact on Dominica as the 'Nature island' tourist destination" (para. 2).

Many comments emerging days after the disaster centered around the loss of vegetation and its lushness which characterized the island domain. Gómez (2017, para. 4) aptly writes "Gusts of wind up to 300 km an hour stripped entire forests, leaving trunks bare, as dry as if they had been sprayed with herbicide. The color green disappeared from the island." Popular nature trails were made permanently impassable and the endemic parrot *Amazona imperialis* and other wildlife were made homeless. Damages to forestry resources were estimated at US$29.7 million including damage to wildlife habitat, nurseries and public information centers (Government of the Commonwealth of Dominica 2017). The hotel sector was not spared with 40 per cent of the room stock incurring severe structural damage and associated tour operators, taxi drivers and ancillary services experiencing financial losses up to US$1.59 million (Government of the Commonwealth of Dominica 2017). As expected, the destination experienced an 88.4% drop in tourist arrivals going into the first half of 2018 (Dominica News Online 2018) with lost government revenue amounting to EC$191 million (US$70.9 million) (Government of the Commonwealth of Dominica 2017).

Tourism Recovery Post Hurricane Maria

Despite the devastating effects on the island and on its tourism sector, Dominica was poised to receive its first cruise visitors Jan 28, 2021, a mere four months after Hurricane Maria (Discover Dominica Authority 2018). The island recorded an increase in arrivals by the end of the fourth quarter of 2018 and a further fifteen per cent increase for the first ten months in 2019 as compared to the same period in 2018. Specifically, an arrival figure of 89, 626 visitors was recorded for 2019 which represented a 42.5 per cent increase in stayover arrivals from 2018, and cruise arrivals also increased by 70 per cent (CTO 2020). In addition, for the winter season period an 88.4% increase was recorded for 2019 compared to the same period in 2018 following Hurricane Maria's passing in late 2017. This level of recovery was heralded as dramatic and significant

(PricewaterhouseCoopers (PWC) 2019) and was also described as a true "Caribbean Tourism Comeback Story" (Britell 2019). PWC (2019, 3) notes that the Citizenship by Investment (CBI) "…has had a major impact on the island's resilience and is likely to remain a significant driver of sustainability."

Methodology

The study targeted public and private sector representatives in Dominica functioning in tourism related sub-sectors, who played a role in tourism recovery actions, post hurricane Maria. Semi-structured interviews were utilized, with the option afforded to respondents to respond to thirteen (13) questions via an online platform or through face-to-face virtual interactions. Specifically, strategic interviews were pursued to elicit tactical insights, feedback and intuitions from key stakeholders in the Dominica tourism sector (Ratcliffe 2002). The request for interviews with the option to directly respond to the questions was disseminated to policy makers, destination managers and tourism practitioners via email. A total of four (4) in-depth responses were received to questions posed (see Table 6.1). Additionally, secondary research was also conducted by way of desktop review of reports on post-hurricane Maria recovery efforts produced by the Government of Dominica, The World Bank, the United Nations Development Programme and the Caribbean Tourism Organization.

Findings and Discussion

Tourism and Disaster Preparedness

The interview guide included one question, which specifically asked about disaster planning in Dominica prior to Hurricane Maria. The responses did not specifically indicate that a comprehensive disaster management plan and protocols existed for the tourism sector prior. Further secondary analysis indicated the existence of a National Disaster Plan 2001 (Government of Dominica 2001) however, it was noted that transference of the plan to the local communities was problematic with the absence of basic actions such as emergency drills. One report stated *"the framework created for these emergency plans is a top-down policy. The actions listed in the plan are to task responsibilities assigned to specific government officials, secretaries,*

Table 6.1 List of questions used for semi-structured interviews

1. What strategy/strategies were used to assist with tourism recovery post Hurricane? For example, hiring an external consultant or developing a national recovery strategy etc. (Please provide as much detail as possible)	2. Was a task force established to assist with tourism recovery post Hurricane Maria? If yes, what was the composition and in your estimation was the task force effective?
3. Were there any protocols (e.g. communication, evacuation, shelters etc.) established by government for stakeholder and visitor management prior to Hurricane Maria? If yes/no, please explain	4. What strategies were used for the restart of the tourism sector and to support tourism development (e.g. incentives, packages, new types of tourism products/experiences etc.) post Hurricane Maria and why?
5. What strategies were used for destination marketing post Hurricane Maria? For example, social media, marketing campaigns etc. Please indicate why a particular strategy was used and its effectiveness?	6. How was the brand "The nature island" maintained post Hurricane Maria given the level of destruction to the destination's tourism assets?
7. Which type of travelers were targeted post Hurricane Maria and why?	8. What would you say was your most successful recovery strategy post Hurricane Maria?
9. What would you say was your least successful recovery strategy post Hurricane Maria?	10. Which tourism subsector in your estimation displayed a high level of resilience and was the first to rebound post Hurricane Maria and why? (e.g. accommodation, food & beverage, sites & attractions, tourism services & travel trade, transportation etc.)
11. Which tourism segment and/niche in your estimation displayed a high level of resilience and was the first to rebound post Hurricane Maria and why? (e.g. cruise, adventure tourism, ecotourism, cultural heritage etc.)	12. What were some of the key lessons you/your organization learnt from the implementation of tourism recovery strategies post Hurricane Maria?
12. Upon reflection, is there anything that could have been done differently to further increase tourist arrivals to the destination in the early period post Hurricane Maria?	

and offices" supporting the views expressed by Fakhruddin et al. (2020). Moreover, respondents tended to equate protocols for the tourism sector in the wider context of national plans with one respondent stating that *"all communities were equipped with emergency shelters."* A Rapid Needs

Assessment for the Tourism Sector also indicated the absence of plans and protocols to deal with visitor related matters and effective engagement of stakeholders.

Notably, in the National Resilience Development Strategy Dominica 2030 (Government of Dominica 2018), the Government of Dominica indicates that they will focus on the incorporation of disaster risk management mechanisms in tourism related facilities, provision of disaster risk management training for tourism workers and capacity building in climate resiliency and disaster preparedness for industry stakeholders. These actions signal a commitment to the country's goal of becoming the first climate resilient country in the world, and represents a demonstrable fortitude that SIDS are believed to possess (Briguglio 1995; Easterly and Kraay 2000; Armstrong and Read 2002; Croes 2006; Baldacchino and Bertram 2009; Philpot, Gray and Stead 2015). Notwithstanding, it is necessary to avoid the perceived top-down approach ascribed to the development and implementation of the 2001 disaster plan. The Government is clearly demonstrating proactive planning and care in their pre-planning efforts for the tourism sector. However, this would not be useful without establishing strong cooperation, collaborative and participatory structures among all stakeholders (Hystad and Keller 2008; Cochrane 2010; Cheer and Lew 2018; Bierman 2018; Sharma et al. 2021).

Tourism Responses and Recovery Efforts

The study's findings reflect several elements contained in Ritchie's (2004) Tourism Crisis and Disaster Management Framework. Respondents pointed to the creation of a Post Maria Task Force and a Crisis Management Taskforce which were both made up of public and private stakeholders. One respondent stated: *"A Post Maria Task Force made up of public and private sector stakeholders played a critical role. This was led by a private sector person. It identified targets and milestones to begin the road to recovery"* while another indicated:

> a Crisis management task force was established with key stakeholders. Heads from major line Ministries which consist of: Ministries of Tourism, Health, Environmental Health, Communication & Works, Agriculture, Finance, Community Development, Foreign Affairs. Also, reps from the Office of Disaster Management, the media, police, Red Cross and heads

> of utility companies. (PPPs). The task force was extremely effective and successful.

The findings of this study demonstrate a phased approach to the recovery effort and one that reflected the Phoenix phase identified by Reddy et al. (2020). Although 90 per cent of the island's vegetation was destroyed, the tourism recovery strategy continued to leverage on the destination's reputation as the "Nature Isle of the Caribbean" inviting persons to engage in "voluntourism" to rebuild natural assets that were destroyed. One respondent shared:

> Voluntourism packages were developed in collaboration with private sector partners and communicated to consumers via various media outlets. The Ministry of Tourism developed a programme dubbed "Voluntourism", where visitors etc were engaged in clean-up and rehabilitation activities." Another respondent stated: "the messaging was refocused on rebuilding rather than enjoying the nature island, until the vegetation and infrastructure was restored to an acceptable level to receive visitors.

Respondents also indicated that an appeal was made to cruise passengers with the destination calling for cruise lines to return as early as January 1, 2018 as a means of generating income for affected stakeholders and exposing travelers to the destination for a limited timeframe so that they could encourage other visitors to visit the destination and be part of the re-building process. One respondent wrote that it was *"…more about the human spirit than the physical aspects of the destination."* It is noteworthy that Dominica employed a strategy of appealing to those regional and international visitors with a keen loyalty to the destination, a similar approach adopted by Israel (Beirman 2009). One respondent aptly stated, *"we had to pull on the emotional heartstrings of the potential traveler…to 'have a heart' and if you do have a heart you will want to support Dominica's recovery, and no better way than to come visit…help us clean the trail* etc.…*"*.

Furthermore, crisis communications appeared to be robust with the use of a digital communications strategy which disseminated daily updates through a dedicated website titled "Dominicaupdates.com" correcting negative perceptions or falsehoods and appealing to the nature lover's passion to aid in the re-building process (Ritchie 2004; Beirman 2018; Zhang et al. 2021). One respondent shared: *"The Ministry of*

Tourism/Marketing Department with the Private Sector (PPP) also posted press releases/videos via*social media platforms with the "Dominica is opened for business"*, while another stated:

> Given the redirection of Government finances, most public communication was done on the digital platforms and through PR. The voluntourism product was developed and that offering led the public communications to raise awareness...It was felt that Travel with meaning [i.e. for a purpose] was trending and that Dominica's particular circumstance warranted such messaging.

One respondent indicated that in 2021 some of the same digital content still receives a high level of viewership when re-shared during the anniversary period:

> A series of image carousels were used to show progress in nature coming back. On anniversaries after Maria this carousel gets recirculated and has great viewership. We also tried to showcase aspects of the human spirit that were intrinsic and natural, and to correlate a regrowth of nature with a strength and resilience of the human spirit.

This is indicative of a continuation of the re-imaging process that began in the Phoenix stage post-Hurricane Maria which continues to support the existing eco-tourism brand while still appealing to the goodwill of visitors towards the destination. This was complemented by other recovery efforts including fiscal incentives from the government for material, furnishings and fixtures. The findings therefore suggest that Dominica's tourism recovery was bolstered by their flexibility, adaptability, creativity and innovation to maintain visitor interest in the destination and to compel them to support the destination even during and after a disaster (Cochrane 2010; Reddy et al. 2020; Zhang et al. 2021).

Building Tourism Resilience

Respondents shared their perspectives on key learnings gained from recovery efforts post-Hurricane Maria for the tourism sector, that could be applied to any future natural disasters. All respondents indicated the need for better plans to be developed for crisis/disaster management at the national level, and for the tourism sector specifically, while one pointed to the need for activation of these plans. The following sentiments were expressed "*The need for disaster management and business continuity plans are critical if the sector is to recover swiftly from future disaster events.*" As Cheer and Lew (2018), Scarpino and Gretzel (2015) and Ricthie (2008) aptly noted, there must be proactive measures taken to build resilience which solicit a clear understanding by policy makers of disaster planning, reduction and readiness and which prompts the creation of implementable measures. The example of Thailand (Beirman 2018) shows how a proactive approach, "a ready-made strategy to employ" strengthens the destination's resilience. One post-Hurricane Maria report also indicated:

> ... the services of experts/consultants may be considered to identify specific needs, gaps and opportunities, conduct market analysis and feasibility studies and map out a comprehensive national tourism development plan and strategy, which should be linked to promoting sustainability, enhancing resiliency and prioritizing community and private sector engagement.

These processes needed to bolster destination preparedness, response and recovery capacity to mitigate and rebound from the impacts of natural hazards and disasters can be built by strengthening the internal operating systems of Dominica in terms of governance, planning and management. Albeit findings suggest that where focus is most needed is in strengthening the collaborative approach with stakeholders (Cochrane 2010) and leveraging on perspective from the "ground" while minimizing bureaucracy Fakhruddin et al. (2020).

Notwithstanding, Fakhruddin et al. (2020) criticism of Dominica's lack of modern information technologies and under-developed communication channels; findings from this study suggest that there was optimal use of information technologies to communicate with tourism markets. Respondents unanimously agree that the most successful recovery strategy was the implementation of a robust digital communications strategies which provided factual information on the state of the island during that period. It is evident that the use of new technologies enhances capacities and capabilities of destinations to generate new products and experiences and to communicate in real time with existing and potential source markets.

One respondent did indicate that communication with locals and organization of relief efforts could have been better executed. In this regard, the recommendations of Cheer and Lew (2018) and Sharma et al. (2021) all emphasize the need for social support systems and the inclusion of the local community. Tourism resilience efforts must include the domestic tourism market and more so ensure that those within the tourism supply chain are kept abreast of developments and overall initiatives via modern technology and communication platforms.

Conclusion

This study exemplifies that there is no single mechanism for building resilience across SIDS or disaster typologies (ECLAC 2011). However, it demonstrates that tourism-dependent islands are not predestined for failure based on inherent vulnerabilities but rather through the adoption of appropriate actions, rapid recovery and robust resilience can both be achieved. Based on the study's findings, Dominica exhibited flexibility and adaptability supported by innovative, creative and technologically driven actions which appealed to travelers and their love for the destination, prompting a relatively swift return to the destination. Lessons learnt from this experience may prove to be useful for Dominica and other SIDS in a post COVID-19 context. Emerging in a post COVID-19 period is a traveler who may be less inclined to engage in long haul travel due to the uncertainties that surround the virus even with the implementation of global vaccination programmes. An appeal to loyalists may be needed in the first instance to stimulate recovery efforts. At the time of this writing, Dominica and other SIDS are in their Phoenix phases of the COVID-19 response and recovery strategy, with many marketing new tourism experiences. For Dominica, the "Safe in Nature" remote work programme encourages individuals to live, work and vacation in nature. This approach has been adopted by many Caribbean SIDS, benefitting from the upsurge in remote work, which has been spurred by physical distancing and containment measures to contain the spread of the COVID-19 virus, and seeking to appeal to the "Digital Nomad" market.

As Dominica forges towards becoming the first climate resilient country in the world with the need to build a more resilient tourism sector, given the dependence on tourism for social and economic development, there must be cognizance of the need for strong planning and implementation, for training and public awareness programmes, and for local community engagement as well as private and public sector collaboration. This must be supported by good governance that utilizes a participatory approach, and investment and use of new information communication technologies.

References

Adger, W.N. 2000. Social and Ecological Resilience: Are They Related? *Progress in HuMan Geography* 24: 347–364.

Armstrong, H.W., and R. Read. 2002. The Phantom of Liberty? Economic Growth and the Vulnerability of Small States. *Journal of International Development* 14 (3): 435–458.

Association of Caribbean States. 2017. Natural Hazards in the Caribbean: Too costly to ignore? http://www.acs-aec.org/index.php?q=disaster-risk-reduction/natural-hazards-in-the-caribbean-too-costly-to-ignore. Accessed February 10, 2021.

Baldacchino, G., and G. Bertram. 2009. The Beak of the Finch: Insights into the Economic Development of Small Economies. *The round Table* 98 (401): 141–160.

Bettencourt, S., R.S. Pryce, and H. Gitay. 2014. Climate Change and Natural Disasters in Small Island Developing States. file:///C:/Users/Tenisha/Downloads/Climate%20change%20and%20natural%20disasters%20(1).pdf. Accessed January 20, 2021.

Betzold, C. 2015. Adapting to Climate Change in Small Island Developing States. *Climatic Change* 133 (3): 481–489.

Beirman, D. 2009. Crisis and Post-Crisis Tourism Destination Recovery Marketing Strategies. In *Crisis Management in the Tourism Industry Beating the Odds?* Edited by Christof Pforr and Peter Hosie, 207–224. New York, NY: Routledge.

Beirman, D. 2018. Thailand's Approach to Destination Resilience: An Historical Perspective of Tourism Resilience from 2002–2018. *Tourism Review International* 2: 227–292.

Briguglio, L. 1995. Small Island Developing States and Their Economic Vulnerabilities. *World Develop* 23 (9): 1615–1632.

Bueno, R., C. Herzfeld, E. A, Stanton, and F. Ackerman. 2008. The Caribbean and Climate Change: The Costs of Inaction. Sweden: Stockholm Environment Institute (SEI) Stockholm.

Britell, A. 2019. Dominica, a Caribbean Tourism Comeback Story. https://www.caribjournal.com/2019/07/17/dominica-caribbean-tourism-story/. Accessed February 4, 2021.

Caribbean Community (CARICOM) Secretariat. 2016. Dominica. https://agricarib.org/country/dominica. Accessed March 2, 2021.

Caribbean Tourism Organization. 2020a. *Multi-Hazard Risk Management Guide for the Caribbean Tourism Sector Climate Change Mitigation and Adaptation Tourism Sector Best-practice Guidelines Handout*. Barbados: CTO.

Caribbean Tourism Organization. 2020b. Multi-Harzard Risk Management for the Caribbean Tourism Sector [Online Course]. Barbados: CTO.

Cheer, J.M., and A.A. Lew. 2018. Understanding Tourism Resilience: Adapting to Social, Political, and Economic Change. In *Tourism, Resilience and Sustainability: Adapting to Social, Political and Economic Change*, ed. J. M. Cheer and A. A. Lew, 3–17. New York, NY: Routledge.

Cochrane, Janet. 2010. The Sphere of Tourism Resilience. *Tourism Recreation Research* 35 (2): 173–185.

Commonwealth Secretariat. 2021. Dominica. https://thecommonwealth.org/our-member-countries/dominica. Accessed March 2, 2021.

Croes, R.R. 2006. A Paradigm Shift to a New Strategy for Small Island Economies: Embracing Demand Side Economics for Value Enhancement and Long-Term Economic Stability. *Tourism Management* 27 (3): 453–465.

Dasgupta, Susmita, B. Laplante, C. Meisner, D. Wheeler, and J. Yan. 2007. *The Impact of Sea Level Rise on Developing Countries: A Comparative Analysis. Policy Research Working Paper; No. 4136*. Washington, DC: World Bank.

Discover Dominica Authority. 2018. Dominica Update - One Year Post Hurricane Maria. https://www.prnewswire.com/news-releases/dominica-update--one-year-post-hurricane-maria-300714744.html. Accessed April 29, 2021.

Discover Dominica Authority. 2021. The History of Dominica. https://discoverdominica.com/en/history-culture. Accessed March 2, 2021.

Dominica New Online. 2018. Officials Say Impact of Hurricanes More Visible in Tourism Industry. https://dominicanewsonline.com/news/homepage/news/general/officials-say-impact-of-hurricanes-more-visible-in-tourism-industry/. Accessed March 2, 2021.

Easterly, W., and A.C. Kraay. 2000. Small States, Small Problems? Income, Growth and Volatility in Small States. *World Development* 28 (11): 2013–2027.

Economic Commission for Latin America and the Caribbean. 2011. *Study on the Vulnerability and Resilience of Caribbean Small Island Developing States (SIDS)*. Washington: ECLAC. https://www.cepal.org/sites/default/files/publication/files/38568/LCCARL354_en.pdf.

Everest-Phillips, M. 2014. Small, So Simple? Complexity in Small Island Developing States. https://www.undp.org/content/dam/undp/library/capacity-development/English/Singapore%20Centre/GPCSE_Complexity%20in%20Small%20Island.pdf. Accessed March 30, 2021.

Faulkner, B. 2001. Towards a Framework for Tourism Disaster Management. *Tourism Management* 22: 135–147.

Fakhruddin, B.S., K. Blanchard, and D. Ragupathy. 2020. Are We There Yet? The Transition from Response to Recovery for the COVID-19 Pandemic. *Progress in Disaster Science* 7: 100–102.

Gómez, S.A. 2017. A Race Against Time to Save Dominica's Biodiversity. http://en.granma.cu/mundo/2017-10-31/a-race-against-time-to-save-dominicas-biodiversity. Accessed March 2, 2021.

Government of the Commonwealth of Dominica. 2001. National Disaster Plan 2001. https://www.preventionweb.net/files/74887_nationaldisasterplan2001dominica.pdf. Accessed April 2, 2021.

Government of the Commonwealth of Dominica. 2017. Post-Disaster Needs Assessment Hurricane Maria September 18, 2017. https://resilientcaribbean.caricom.org/wp-content/uploads/2017/11/DOMINICA-EXECUTIVE-SUMMARY.pdf. Accessed March 15, 2021.

Government of the Commonwealth of Dominica. 2018. The National Resilience Development Strategy - Dominica 2030. https://dominica.gov.dm/images/documents/national_resilience_development_strategy_2030.pdf. Accessed April 2, 2021.

Gurtner, Y. 2016. Returning to Paradise: Investigating Issues of Tourism Crisis and Disaster Recovery on the Island of Bali. *Journal of Hospitality and Tourism Management* 28: 11–19. https://doi.org/10.1016/j.jhtm.2016.04.007.

Haimes, Y.Y. 2009. On the Definition of Resilience in Systems. *Risk Analysis: An International Journal* 29 (4): 498–501.

Holladay, P.J., and R.B. Powell. 2016. Social-Ecological Systems, Stakeholders and Sustainability: A Qualitative Inquiry into Community-Based Tourism in the Commonwealth of Dominica. *Caribbean Studies* 44 (1–2): 3–28.

Hosseini, S., K. Barker, and J.E. Ramirez-Marquez. 2016. A Review of Definitions and Measures of System Resilience. *Reliability Engineering & System Safety* 145: 47–61.

Hystad, P.W., and P.C. Keller. 2008. Towards a Destination Tourism Disaster Management Framework: Long-Term Lessons from a Forest Fire Disaster. *Tourism Management* 29 (1): 151–162.

IUCN. 2008. One Third of Reef-Building Corals Face Extinction. https://www.iucn.org/ru/node/1951. Accessed April 20, 2021.

Invest Dominica Authority. 2021. Economic Statistics. http://www.investdominica.com/the-ida-full-guide-to-doing-business-in-dominica/economic-statistics/. Accessed March 15, 2021.

Khazai, B., F. Mahdavian, and S. Platt. 2018. Tourism Recovery Scorecard (TOURS)–Benchmarking and Monitoring Progress on Disaster Recovery in Tourism Destinations. *International Journal of Disaster Risk Reduction* 27: 75–84.

Lieberman, A. 2020. Dominica Determined to Become world's First Climate-Resilient Nation. https://www.devex.com/news/dominica-determined-to-become-world-s-first-climate-resilient-nation-96999. Accessed March 15, 2021.

Luthe, T., and R. Wyss. 2014. Assessing and Planning Resilience in Tourism. *Tourism Management* 44: 161–163.

Mair, J., B.W. Ritchie, and G. Walters. 2016. Towards a Research Agenda for Post-Disaster and Post-Crisis Recovery Strategies for Tourist Destinations: A Narrative Review. *Current Issues in Tourism* 19 (1): 1–26. https://doi.org/10.1080/13683500.2014.932758.

Mooney, H., and M.A. Zegarra. 2020. Extreme Outlier: The Pandemic's Unprecedented Shock to Tourism in Latin America and the Caribbean. https://servicetickets.us/files/documents/Extreme-Outlier-The-Pandemics-Shock-to-Tourism.pdf. Accessed March 15, 2021.

National Weather Service. 2017. Major Hurricane Maria - September 20, 2017. https://www.weather.gov/sju/maria2017#:~:text=Maria%20formed%20from%20an%20African,to%20September%2017th%2C%202017. Accessed April 27,2021.

OECD. 2018. *Making Development Co-operation Work for Small Island Developing States*. Paris: OECD Publishing.

Office of the High Representative for the Least Developed Countries, Landlocked Developing Countries and Small Island Developing States (UN-OHRLLS). 2011. Small Island Developing States Small Islands Big(Ger) Stakes. http://unohrlls.org/custom-content/uploads/2013/08/SIDS-Small-Islands-Bigger-Stakes.pdf. Accessed December 11, 2019.

Ötker, I., and Srinivasan, K. 2018. Bracing for the Storm. https://www.imf.org/external/pubs/ft/fandd/2018/03/otker.htm. Accessed March 15, 2021.

Pacific Community (SPC). 2016. Framework for Resilient Development in the Pacific An Integrated Approach to Address Climate Change and Disaster Risk Management (FRDP) 2017 – 2030. http://gsd.spc.int/frdp/assets/FRDP_2016_Resilient_Dev_pacific.pdf. Accessed March 15, 2021.

Philpot, D., T.S. Gray, and S.M. Stead. 2015. Seychelles, a Vulnerable or Resilient SIDS? A Local Perspective. *Island Studies Journal* 10 (1): 31–48.

PricewaterHouse Coopers. 2019. Assessment of the Commonwealth of Dominica's Citizenship by Investment Programme. https://cbiu.gov.dm/wp-content/uploads/2019/08/PwC-Report_-Dominica-CBI_19-August.pdf. Accessed March 15, 2021.

Ratcliffe, J. 2002. Scenario Planning: Strategic Interviews and Conversations. *Foresight* 4 (1): 19–30. https://doi.org/10.1108/14636680210425228.

Reddy, M.V., S.W. Boyd, and M. Nica. 2020. Towards a Post-Conflict Tourism Recovery Framework. *Annals of Tourism Research*. https://doi.org/10.1016/j.annals.2020.102940.

Ritchie, B.W. 2004. Chaos, Crises and Disasters: A Strategic Approach to Crisis Management in the Tourism Industry. *Tourism Management* 25 (6): 669–683.

Ritchie, B.W. 2008. Tourism Disaster Planning and Management: From Response and Recovery to Reduction and Readiness. *Current Issues in Tourism* 11 (4): 315–348. https://doi.org/10.1080/13683500802140372.

Ritchie, B.W. and Y. Jiang. 2019. A Review of Research on Tourism Risk, Crisis and Disaster Management: Launching the Annals of Tourism Research Curated Collection on Tourism Risk, Crisis and Disaster Management. *Annals of Tourism Research* 79: 102812.

Rose, A. 2004. Defining and Measuring Economic Resilience to Disasters. *Disaster Prevention and Management* 13 (4): 307–314.

Scarpino, M., and U. Gretzel. 2015. Conceptualizing Organizational Resilience in Tourism Crisis Management. In *Tourism Crisis and Disaster Management in the Asia Pacific*, ed. B. Ritchie and K. Campiranon, 15–32. UK: CABI Publishing.

Sharma, G.D., A. Thomas, and J. Paul. 2021. Reviving Tourism Industry Post-COVID-19: A Resilience-Based Framework. *Tourism Management Perspectives*. https://doi.org/10.1016/j.tmp.2020.100786.

Simmie, James, and Ron Martin. 2010. The Economic Resilience of Regions: Towards an Evolutionary Approach. *Cambridge Journal of Regions, Economy and Society* 3 (1): 27–43.

Simpson, M., S. Gössling, and D. Scott. 2008. *The International Policy and Market Response to Global Warming and the Challenges and Opportunities that Climate Change Issues present for the Caribbean Tourism Sector*. Barbados: Caribbean Tourism Organization.

Sjöstedt, M., and M. Povitkina. 2017. Vulnerability of Small Island Developing States to Natural Disasters: How Much Difference Can Effective Governments Make? *The Journal of Environment & Development* 26 (1): 82–105.

Slany, A. 2020. Multiple Disasters and Debt Sustainability in Small Island Developing States. In *United Nations Conference on Trade and Development Research Paper No. 55*, edited by UNCTAD, 1–56. https://unctad.org/system/files/official-document/ser-rp-2020d14_en.pdf.

Thomas, A., and S. Lindo. 2019. *Disappearing Islands: Small Island Developing States and Climate Change. Environment Issues: Science, Policy, and Diplomacy*. San Diego: Cognella.

Thomas A., A. Baptiste, R. Martyr-Koller, P. Pringle, and K. Rhiney. 2020. Annual Review of Environment and Resources Climate Change and Small Island Developing States. https://www.annualreviews.org/doi/pdf/10.1146/annurev-environ-012320-083355. Accessed April 12, 2021.

UNCTAD. 2021. For heavily indebted small islands, resilience-building is the best antidote. https://unctad.org/news/heavily-indebted-small-islands-resilience-building-best-antidote. Accessed March 20, 2021.

UNEP .2014. Emerging Issues for Small Island Developing States. https://sustainabledevelopment.un.org/content/documents/2173emerging%20issues%20of%20sids.pdf. Accessed January 20, 2021.

United Nations World Tourism Organization. 2020. *UNWTO Briefing Note – Tourism and COVID-19, Issue 2. Tourism in SIDS – the challenge of sustaining livelihoods in times of COVID-19*. Madrid: UNWTO.

Weis, J. 2021. Apocalyptic' St. Vincent Eruption May Cost Up to 50% of GDP. https://www.bloomberg.com/news/articles/2021-04-22/-apocalyptic-st-vincent-eruption-may-cost-up-to-50-of-gdp. Accessed April 29, 2021.

Wetzel, F.T., W.D. Kissling, H. Beissmann, and D.J. Penn. 2012. Future Climate Change Driven Sea Level Rise: Secondary Consequences from Human Displacement for Island Biodiversity. *Global Change Biology* 18 (9): 2707–2719.

World Bank. 2017. A 360 degree look at Dominica Post Hurricane Maria. https://www.worldbank.org/en/news/feature/2017/11/28/a-360-degree-look-at-dominica-post-hurricane-maria. Accessed April 12, 2021.

World Bank. 2020. *Rebuilding Tourism Competitiveness: Tourism Response, Recovery and Resilience to the COVID-19 Crisis*. Washington, DC: World Bank.

Zhang, H., H. Song, L. Wen, & C. Liu. 2021. Forecasting Tourism Recovery Amid COVID-19. Annals of Tourism Research 87: 103149.

PART III

Socio-Cultural Resilience

CHAPTER 7

Community Resilience in the Face of Natural Disaster: Puerto Rico's Adventure Tourism Industry

Mechelle N. Best and José H. González

Introduction

The Caribbean Tourism Organization (CTO) promoted 2017 as the "Year of Adventure in the Caribbean" (CTO 2017) and in the realm of adventure tourism, Puerto Rico stands as one of the top contenders in the region. The year 2017 could also be dubbed the "year of natural disasters" with Puerto Rico once again rising to the top. On September 20, 2017, Hurricane Maria, a nearly Category 5 hurricane, made a direct hit on Puerto Rico causing catastrophic damage and leaving many parts of the inhabited islands non-operational or close to non-operational for months after the hurricane made landfall. The devastation caused by

M. N. Best · J. H. González (✉)
Department of Recreation and Tourism Management, California State University Northridge, Los Angeles, CA, USA
e-mail: jose.gonzalez@csun.edu

M. N. Best
e-mail: mechelle.best@csun.edu

A. Lewis-Cameron et al. (eds.), *Managing Crises in Tourism*,
https://doi.org/10.1007/978-3-030-80238-7_7

Maria affected many sectors, including tourism as a whole and adventure tourism in particular. Hurricane Maria, tested the resilience of Puerto Rico's adventure tourism community, but the community was up to the challenge.

This chapter explores the resilience of Puerto Rico's adventure tourism community in light of the worst natural disaster experienced by the Commonwealth of Puerto Rico in recorded history. Data collected through interviews with owners and operators of adventure tour companies and other industry experts in Puerto Rico in late summer of 2019 provide the foundation for the adventure tourism community discussion in this chapter. The tour companies featured facilitate hiking, zip-lining and other adventure activities in El Yunque National Forest and other natural locations on the main island. The study objectives reported in this chapter are:

1. To explore the factors that contribute to the community resilience of Puerto Rico's adventure tourism industry.
2. To determine if a natural disaster can be a catalyst for change in Puerto Rico's adventure tourism industry.

Contextualizing Resilience, Natural Disasters, and Adventure Tourism

Resilience, according to Holling (1973, 14) is "a measure of the persistence of systems and their ability to absorb change and disturbance and still maintain the same relationships between populations or state variables." This concept of resilience was developed for ecological systems and has over the decades since been applied to social and economic systems as well, particularly in the context of social impacts from natural hazards and disasters where the focus tends to be on how quickly social systems or industry sectors can return to normal in the aftermath of a disturbance of the system (Biggs 2011; Biggs et al. 2012).

The United Nations Office for Disaster Risk Reduction (UNDRR 2017, para. 1) defines resilience as:

> The ability of a system, community or society exposed to hazards to resist, absorb, accommodate, adapt to, transform and recover from the effects of a hazard in a timely and efficient manner, including through the preservation

> and restoration of its essential basic structures and functions through risk management.

In the tourism context, Holladay and Powell (2016, 9) propose that resilience is "the amount of change a system ... can undergo and retain its same function, structure, and feedbacks."

Community Resilience

A community may be tied together through location, values, culture, religion, customs, or other factors. Despite commonalities, community members are heterogeneous which complicates the process of determining what strategies are optimal to make a specific community more resilient (Twigg 2009). Residents drive resilience in some communities; likewise, businesses in others. In their research on community resilience, Cutter et al. (2008, 599) define resilience as.

> The ability of a social system to respond and recover from disasters and includes those inherent conditions that allow the system to absorb impacts and cope with an event, as well as post-event, adaptive processes that facilitate the ability of the social system to re-organize, change, and learn in response to a threat.

Twigg (2009), in his guidance note, *Characteristics of a resilient community*, laid out a resource which can be used by community stakeholders to work collectively to minimize their disaster risk. Recognizing that communities are not homogenous, the author created a resource that communities can modify to suit their unique situations. The resource offers a wide-ranging list of characteristics within five thematic areas: governance; risk assessment; knowledge and education; risk management and vulnerability reduction; and disaster preparedness and response. The characteristics include:

- *Community disaster plans publicly available and widely understood.*
- *All sections of community know about facilities/services/skills available pre-, during and post-emergency, and how to access these.*
- *Community understanding of characteristics and functioning of local natural environment and ecosystems (e.g. drainage, watersheds, slope,*

and soil characteristics) and the potential risks associated with these natural features and human interventions that affect them.
- *Small enterprises have business protection and continuity/ recovery plans*. (Twigg 2009, 36–40).

Tourism destinations are communities with both residents and businesses along with other stakeholders and resources that contribute to their resilience. Smaller groupings within communities (e.g. the adventure tourism sector) also constitute communities in their own right. Holladay and Powell (2016) in their study of the resilience of sustainable tourism at the community level in the Commonwealth of Dominica, found that local communities need to be empowered to be self-reliant and self-organized. Without these changes, community resilience would falter because decision making was generally top-down from the central government.

Natural resources also play a role in a community's resilience. Researchers have proposed that community resilience is interconnected with the natural systems within and around it and that a degraded environment increases a community's vulnerability to natural hazards (Cutter et al. 2008). Similarly, recovery of the natural environment after a disaster can help the community to recover as well. The community can also aid in the recovery of the natural environment, and through appropriate use and conservation, can help to make it more resilient to natural hazards (Simmons 2018).

Resilience in Tourism Businesses

Businesses, while part of communities, have their own specific challenges with resilience. Resilience for a business enterprise speaks to its ability to survive a hazard, whether natural or manmade, rebound, and perhaps grow (Dahles and Susilowati 2015). Resilience in this context is sometimes exemplified though an enterprise's innovation, reorganization, and/or adaptation in response to a disruption (Biggs 2011). For example, Biggs, Hall and Stoeckl (2012, 647) define a resilient reef tourism enterprise as "one that is able to maintain or grow its existing level of employment and income and stay operating in reef tourism in the face of one or more shocks or crises."

Various factors contribute to a business' resilience. These include: size; human capital; owners' lifestyle values (e.g. love for working in natural settings); health of the natural resources or ecosystems; access to financial capital; support from government or non-governmental organizations; and intra-industry support (Biggs 2011; Biggs et al. 2012). In their study of Mississippi small businesses' resilience to Hurricane Katrina, Torres, Marshall, and Sydnor (2019) found that social capital is a key asset for long-term resilience. They also found that small business owners who were connected to their communities were more likely to overcome disaster and build resilience. Furthermore, the more links business owners have to the community, the more social capital they have. This is critical because self-reliance alone cannot assure long-term post disaster recovery and resilience.

Natural Disasters

Natural disasters are common phenomena in the Caribbean, particularly disasters ensuing from hurricane strikes. The Caribbean Disaster Emergency Management Agency (2018, 1) noted that with its history of volcanic eruptions, earthquakes, hurricanes, and floods the Caribbean has come to be regarded as the "second most disaster prone region in the world." With the Caribbean's dependence on tourism, there is additional cause for concern about hurricanes and other natural hazards. Hurricanes can also have significant impacts to the natural environment, even in those instances where no human lives are lost and structural damage is minimal or moderate. Conversely, disasters also act as catalysts for positive changes in natural and manmade systems (Henderson 2005; Johnson 2015).

Natural disasters are also catalysts for change within businesses and in the business community. According to Rindrasih (2019), the tsunami of 2004 opened a new window of opportunity for tourism in Aceh, Indonesia by providing the impetus for a peace agreement after decades of political conflict. Rindrasih (2019) further suggests that disasters may provide windows of opportunity for tourism development encourage a destination to "build back better," using more local input and therefore fomenting community "buy-in" for proposed tourism development, leading to increased resilience of the destination.

Adventure Tourism

Adventure is defined as "any activity or journey that creates a sense of risk and thrill for the person participating in an activity which may have a degree of risk of injury" while adventure tourism is tourism that involves "an activity that provides the participant with a degree of 'perceived risk' outside of their normal place of residence" (Jenkins 2019, 4).

Context for the Study

The Commonwealth of Puerto Rico is an archipelago of over 140 islands, cays and islets in the northern Caribbean. It has been a US territory since 1898. The main island, Puerto Rico, two smaller islands Vieques and Culebra, and the islet Old San Juan are the only inhabited ones in the archipelago. Puerto Rico has a plethora of natural resources including lakes, dry forest, rainforest, and an extensive coastal and marine ecosystem and is one of the Caribbean's best known tourism destinations.

Puerto Rico has many adventure activities to exceed the expectations of adventurists. They include hiking in El Yunque National Forest, paddling to mangrove keys, surfing Gas Chambers at "The Caribbean Pipeline," and enjoying a zipline/canopy experience. These types of adventure activities in Puerto Rico can be traced back to the YMCA's zipline which opened in 1978/79, the first such in Puerto Rico (L. Acevedo 2019, pers. comm.). Despite its history of outdoor and adventure types of recreation, it was not until 1997 that Aventura's Tierra Adentro, the first registered company offering caving tours, was established in Puerto Rico. It would be a few more years before another company, Acampa, started offering nature trips and promoting Puerto Rico as a destination for adventure tourism (R. Sepulveda 2019, pers. comm.).

With these and other adventure activities all dependent on natural resources, the adventure tourism industry is particularly vulnerable to natural disasters. Its vulnerability is further heightened by dependence on the US mainland as its primary source market. Tourist arrivals (stay-over and excursionists) to Puerto Rico grew from 4.21 million per year in 2013, to some 5.19 million in 2018. During that 5-year period, 85–87 percent of visitors were from the mainland (UNWTO 2018). Anecdotally it is known that many tourists enjoy adventure activities, though specific data are not collected to provide a more detailed understanding of the proportion that participates.

Hurricane Maria, September 2017

Hurricane Maria was the 13th named storm of the 2017 Atlantic hurricane season and was the first category 4 hurricane to hit Puerto Rico since 1928 (Office for Coastal Management [OCM] 2019). On September 20th, Maria made landfall on Puerto Rico with wind speed of around 155 mph, just under the threshold of a category 5 (Pasch et al. 2019). Maria made landfall a mere two weeks after Hurricane Irma skirted the northern part of the main island and with its direct strike, compounded Irma's damage while wrecking tremendous additional damage of its own (Meyer 2017).

There was extensive infrastructural damage, including to the tourism industry: power outages for the majority; fuel shortages; loss of cell service, Internet and telephone landlines; limited potable water; (Meyer 2017; OCM 2019; US DoD 2017). Additionally, the main airport, Luis Muñoz Marín International in San Juan, and seven other small ones were open with restrictions and primarily only during daylight, while the Port of San Juan and seven other seaports were open with restrictions (US DoD 2017). HurricaneMaria's damage to Puerto Rico's tourism industry was severe. The World Travel and Tourism Council (2018a, b) estimated that following Maria, Puerto Rico lost around 202,000 visitors, US$133.4 million in visitor spend, US$43.5 million in direct contribution to GDP and 500 jobs.The hotel sector was significantly impacted with closures ranging from one month to indefinitely. It was estimated that 2,556 rooms or 16% of the hotel room stock were lost (WTTC 2018a, b). However, this figure does not account for hotels that may have remained standing, but were non-operational. Hurricane Maria caused major short- and long-term impacts to Puerto Rico's natural environment, including to the wildlife (terrestrial and marine) and vegetation on which adventure tourism depends. Reports initially suggested that El Yunque National Forest was decimated. Indeed, the National Forest, other forests, and trees across the island were hard hit but many survived (Rosenbloom 2017).

Methodology

This exploratory investigation of the resilience of Puerto Rico's adventure tourism industry stemmed from the researchers' interests in adventure recreation and sustainable tourism and deep concern about how the community of the adventure tourism niche was impacted by Hurricane Maria. Before Hurricane Maria, there were 39 nature and adventure companies (including 14 zip-line tour companies) listed in *Que Pasa*, Puerto Rico's official tourism magazine at the time (Casiano Communications n.d). These companies along with other tourism industry components, comprise the adventure tourism industry and constitute a "community" with a common interest of offering adventure recreation to locals and tourists.

Given the extensive impacts from Hurricane Maria and the fact that this is a relatively nascent area of research in Puerto Rico, it was determined that in-person semi-structured interviews of select nature/adventure tour operators and tourism and adventure experts were appropriate to conduct a pilot for a broader study. Ten potential participants were short-listed based on annual visitation rate, scope of practice, years of service, and areas of expertise. Participants were invited by email and 7 accepted.

Interviews were completed in Puerto Rico during a one-week period in August 2019. Each interview lasted approximately two hours and they were conducted on the site of the operation. Common base questions were used to guide each interview. Conducting in-person interviews at the adventure sites where possible, allowed for observation of the sites' recovery first-hand and comparison between current state and pre- and post-Maria photos and videos provided by the operators. While no one from the National Forest Service was interviewed, the researchers visited El Yunque National Forest to observe the current state of the facilities and natural resources and to hike one of the open trails. The visitors center also offered additional photographs, maps, and informational material. The researchers conducted all interviews together and each interview was recorded. Initially, they individually analyzed the interviews, then together reviewed and agreed on their findings.

Adventure Tour Operators

In 2019, only six companies of the aforementioned 39 companies were still operating. This chapter features four of the six: Acampa, Paddle Paradise & Outdoor Center, EcoQuest Adventures & Tours, and ToroVerde Nature Adventure Park.

Acampa opened in 1998 as an independent outdoor retail store and adventure tours started in 2001. Before Hurricane Maria, Acampa had two trekking and zipline sites in San Lorenzo and Toro Negro, operated by 20 employees. Maria destroyed both sites. With the estimated cost to repair them and restart operation upwards of US$200,000 and insurance to run the tours up fivefold, the owner, Raymond Sepulveda, decided against reopening them. Nonetheless, Acampa as a business proved resilient and the company reemerged with a change in focus for its adventure activities, eventually offering exclusive expeditions to Isla de Mona (Mona Island), "the Galapagos" of the Caribbean, instead of offering zipline experiences (R. Sepulveda 2019, pers. comm).

Paddle Paradise PR and Outdoor Center was established in 2013 by Omar and Jessica Ramos and was still a relatively young business when Hurricane Maria struck in 2017. The company offers kayaking and stand-up paddleboarding tours on the water fronting their property as well as ziplining, camping, and team building programmes. Prior to Hurricane Maria the company had 8 staff members and was a solid business. Maria caused the loss of several trees and blanketed the property with debris. It took the owners some 20 days to get back into their facility and they were only able to do so because of their community's efforts. Twenty-two truckloads of natural debris and about 17 truckloads of trash were removed from the property. The cleanup cost alone was about $30,000 and was all out-of-pocket since the insurance was slow to process. Two years after the hurricane, they were still waiting on the insurance. After Maria they lost their employees but rehired them once the business reopened (O. Ramos 2019, pers. comm.).

EcoQuest Adventures and Tours opened in 2004 and is owned and operated by Ivan Purcell. The company operates a zipline course at Campo Rico (a site shared by a few businesses) which suffered less physical damage than other adventure sites around the island. After Maria it took 3 weeks of cleanup to get rid of the debris. The zip-lines had remained intact and Purcell replaced only one cable as a precaution rather than a necessity. EcoQuest's operations restarted in November 2017 because the businesses at Campo Rico worked together to clean up the site. They also supported their community's recovery efforts (I. Purcell 2019, pers. comm.).

ToroVerde Nature Adventure Park started operating in 2008, when ziplining was still just a small segment of the adventure industry in Puerto Rico and in 2016, "The Monster" their longest zipline, was the longest in the world (PRTC 2019). Maria caused a six-month closure of ToroVerde despite minimal infrastructural or natural resource damage on site. Most roads in the area around ToroVerde were closed, limiting access. They were also without electricity for that six-month period. Some employees stayed on during the closure, others returned for reopening (J. Jorge and S. Rivera 2019, pers. comm.).

Community Resilience in Puerto Rico's Adventure Tourism Industry.

Based on the researchers' individual and joint analysis, the themes discussed in the following sections were determined to be the most salient areas of convergence under the overarching community resilience umbrella. Puerto Rico's adventure tourism community exhibited community resilience, built on the resilience of individual businesses, natural resource resilience, and institutional support. While the tour operators put significant emphasis on what they could do independently and strategized to strengthen their capacity, it was also because of the efforts of various stakeholders in the community that the industry rebounded in the aftermath of Hurricane Maria.

Resilient Businesses

Resilient community members drive community resilience. Acampa, Paddle Paradise and Outdoor Center, EcoQuest Adventures and Tours, and ToroVerde Nature Adventure Park each exhibit characteristics of resilient businesses, in keeping with Dahles and Susilowati's (2015) description of what constitutes a resilient enterprise: the ability to survive, rebound and grow where possible. They survived Hurricane Maria and in the ensuing two years made significant progress in rebuilding or expanding in innovative ways, changing their business model to strengthen their businesses. For example, Raymond Sepulveda initially refocused on the outdoor retail side of Acampa before expanding to offer excursions to Mona Island. Ian Purcell expanded EcoQuest by acquiring another adventure company, Yunque Ziplining, that did not survive Maria. Paddle Paradise also adopted a new business model with a new name and brand, to offer a broader spectrum of services.

In addition to the aforementioned changes, the tour operators also focused more on the local market to help with recovery. With travel from the mainland and international markets significantly reduced, the tour operators adapted their programs (e.g. more weekend programming and different activity packages) and prices to be more attractive to local recreationists. Biggs (2011) noted that resilient businesses innovate, reorganize, and adapt when faced with disruption. These tour operators were able to pivot and do the same.

The resilience of the adventure companies was also buttressed by their social and financial capitals. The four companies ranged from smaller-scale (Acampa, Paddle Paradise) to medium (EcoQuest), to large (ToroVerde). From the way in which the small and medium operators discussed their businesses it was clear that their individual love for nature and adventure recreation propelled their entry to the industry and help to explain why despite the difficulties endured to recover and rebuild after Maria, they persisted. Their intense love of Puerto Rico's natural environment and the opportunity to share nature with others drive their commitment. While they each want their business to be successful, they are not motivated by profits alone because they have other connections to the sites and natural resources central to their operations. Biggs et al. (2012) label this as a "lifestyle benefit" which contributes to a business' resiliency. Entrepreneurs like Sepulveda, Ramos, and Purcell are considered "lifestyle entrepreneurs" and lifestyle entrepreneurship is part of their social capital.

Social capital helps to make businesses resilient and stems from the "connections/network between individuals and the level of trust, norms of reciprocity, and cooperation they maintain" (Duffy et al. 2016, 208). The small and medium adventure tour operators belong to networks that supported them in the aftermath of Maria; the owners themselves also supported others within their networks. Family, friends, and community members supported the tour operators and they in turn supported friends and community members. For example, volunteers from the US mainland travelled to Puerto Rico and camped at Paddle Paradise to help them clean up. Additionally, Paddle Paradise was used as the base for storing vehicles and equipment deployed to clear roads in the surrounding community, exemplifying the reciprocity and cooperation integral to social capital. These findings also align with Torres, Marshall, and Sydnor's (2019) conclusion that social capital is an important asset if small businesses are to remain resilient in the long-term and that community support, rather than self-reliance alone makes long-term resilience more likely.

Access to financial capital can determine whether or not a business is able to survive a natural disaster, especially if insurance pay outs aren't immediately forthcoming, as was the situation after Hurricane Maria.

Financial capital was pivotal to the adventure tour operators' recovery. They tapped into personal savings, business loans, support and loans from family and friends, and in one instance, a "GofundMe" campaign started by overseas supporters. Some adventure companies were not as resilient as this group and ceased operating, while others were closed for up to six months. Biggs et al. (2012) in researching the resilience of tourism enterprises in Phuket, Thailand after the 2004 tsunami also found that financial capital (aid or loans from the government, second jobs to supplement business income) helped businesses to survive and recover.

Citing Kates (1977), Sydnor-Bousso et al. (2011, 210–211) note that disasters "rather than fundamentally changing existing social and economic trends in places, simply magnifies trends … if an area is economically distressed pre-disaster, that distress is almost certain to be exacerbated post-disaster." The owner of Acampa had made a significant investment in the business in the months prior to Maria. Consequently, in Maria's aftermath the business had no reserves to tap into. While this is not an example of pre-disaster economic distress, it does highlight that a business' financial situation prior to a hazard influences its post disaster viability. The investment was intended to strengthen the business, but it also weakened its financial capital, rendering it less equipped to tackle the problems resulting from Maria.

An element of resilience is surpassing what the business did before the disaster, not simply rebounding to pre-disaster levels of achievement (Dahles and Susilowati's 2015). Prior to Hurricane Maria, ToroVerde was in a good place financially, it had strong "financial capital." Though it took a few months to reopen after the hurricane, the company was still able to forge ahead with its expansion plans after the hurricane, constructing a new kids' park and continuing to plan for a 50-acre agritourism project and a petting zoo.

These adventure tour operators enhance their community's resilience because they are adaptable, willing to take risks and have strong financial and social capitals to support them. The tour operators are also concerned with strengthening the industry and their vision for adventure recreation in Puerto Rico was undiminished despite Maria's impact. Their confidence in the industry's resilience and ability to continue to adapt and thrive is reflected in their bright outlook for the future. Their vision encompasses the following:

- "Going back to nature"—more streamlined natural resource focused experiences and more private campgrounds due to the closure of public campgrounds and significant usage of Airbnb (Sepulveda).
- Growth in hiking, kayaking, and climbing groups (Ramos).
- More scheduled flights and an uptick in arrivals; DIY (do-it-yourself) adventures, where tourists don't depend on a tour operator, but could easily curate their own adventures with Airbnb properties and ride share services (Purcell).
- "People are looking for an experience not a tour anymore" (Jorge pers. Comm.).

These envisioned changes could also add to the community's resilience through diversification of the adventure tourism product.

Resilient Natural Resources

The resilience of Puerto Rico's adventure tourism community relies on the resilience of natural resources. The community cannot exist without them. Two years after Maria, Puerto Rico's natural resources were recovering well. In response to how quickly some forest areas started to regrow Findlay (2018, para. 10) remarked "for many observing the island's recovery, Maria offered one big lesson: When disaster strikes, it is nature itself that holds the best lessons on resilience." El Yunque National Forest had the majority of the hiking trails open 20 months after Hurricane Maria hit Puerto Rico. In late summer 2019, the researchers observed that several parking areas and the main visitors center remained closed, though a temporary center operated in the town of El Palmar, near the entrance to the rainforest. A 2018 US Forest Service survey showed that though canopy and foliage were lost because of Maria, around 87 percent of tree trunks had withstood the hurricane and remained completely or partially rooted. This increases hope for almost full recovery in the short-term, since evidence from previous destructive hurricanes such as Hugo showed that forests can recover in relatively short time frames of one to five years (Munroe et al. 2018).

Institutional Structures

Institutional structures can strengthen a community's resilience or hinder its ability to recover after a disaster and this was highlighted in the adventure tourism community, pertaining to destination marketing organizations (DMOs), government tourism departments and insurance. DMOs market the destination as a whole with the expectation that benefits will accrue to all stakeholders in the industry. For a tourism destination to recover after a disaster it has to be able to show the market that it is still open for business and oftentimes that with its value proposition it remains superior to its competitors. Therefore, a destination marketing plan with strategies for pre- and post-natural disasters goes a long way in a destination's recovery. One tour operator credited the work done by Discover Puerto Rico (the relatively new DMO) for both encouraging previous visitors to return in the two years after Maria and attracting new visitors, which benefited the adventure tourism community.

The Secretary of Tourism changed three times between 2017 and 2019 bringing instability to the Department of Tourism. In addition, lack of communication and alignment between the Department of Natural Resources and the Department of Tourism made it difficult for adventure tour companies to obtain permits to move forward. In this instance, rather than facilitating the recovery of the community, these institutional structures proved to be barriers. The tour operators deemed insurance to be the biggest barrier to reopening after the Hurricane Maria and to obtaining support from the Department of Tourism for marketing and promotion. The insurance costs for adventure tourism in Puerto Rico doubled, tripled, and even increased seven-fold post Maria. Some businesses simply could not afford insurance and gave up; others operated without the activities that required costly insurance.

Despite the tour operators' positive view of the future, the permanent closure of many adventure tour operator businesses as a consequence of Hurricane Maria and the permitting and insurance challenges, highlights the community's vulnerability. The tour operators support each other, but there is no formal co-operative. While they constitute a community because of their shared interests and informal network, the absence of a formal community means that they deal individually with the aforementioned challenges and many more. In this regard they could benefit from creating institutional structures of their own.

The lack of a formal grouping also means that it is difficult for the community as a whole to implement certain strategies to minimize disaster risks and augment the community's resilience, like those suggested by Twigg (2009). For example, through a formal co-op they could develop disaster plans specific to their niche and as a collective be better positioned to attract funding through innovative mechanisms. With a more formal arrangement the tour operators would be able to approach common challenges as a collective, yet accrue benefits for the individual companies. Holladay and Powell (2016) argued that local communities needed to be self-reliant and self-organized. This is apropos for Puerto Rico's adventure tourism industry. This lack of structure to the community weakens its resilience.

Disasters as Catalysts for Change

In the natural world, fire is necessary for some types of seeds to open and germinate. In the business world, disasters can also be catalysts for innovation, change, and growth. In discussing post-disaster responses to the 2004 tsunami in Phuket, Henderson (2005) noted positive changes spurred by the disaster: sand on the beach seemed finer, while seawater seemed clearer. The clarity of the seawater made it easier to see the marine ecosystem.These changes would all make a destination more attractive to the market. Similarly, in Puerto Rico, changes to the natural resource base forced some adventure operators to focus on natural resources they had not previously used, developing new products to appeal differently to their market or attract new markets.

For three of the tour operators, Hurricane Maria was also a catalyst for change in other ways. For Paddle Paradise, it was a nudge to expand the teambuilding component of the business. For EcoQuest, it was the opportunity to acquire an existing business that was not sufficiently resilient to recover from Maria. For Acampa, it was a push to refocus on the retail store and to create a new expedition package to Mona Island—distinguishing itself from the rest of the market.

Adventure Tourism Outlook and Closing Thoughts

The components of the pilot study presented in this chapter were guided by two objectives: to explore the factors that contribute to the community resilience of Puerto Rico's adventure tourism industry and to determine if a natural disaster could be a catalyst for change for that community. Business resilience, resilient natural resources, and institutional structures were seen to be significant contributors to community resilience in adventure tourism. Likewise, the responses from the adventure tour operators indicate that a natural disaster—Hurricane Maria—was indeed a catalyst for several critical changes to their business models.

Twigg (2009, 8) posited community resilience as "the capacity to anticipate, minimize, and absorb potential stresses" and further suggested that such resilience also rested on the community's ability to recover or "bounce back" after an event. The adventure tourism industry in Puerto Rico, though changed by Hurricane Maria, generally demonstrated community resilience. It is noteworthy that while the community anticipated impacts from Maria, the reality was still vastly beyond their expectations. The tour operators and experts had experienced hurricanes prior to Maria, but none had experienced any on the scale of Maria and the destruction it wrought. The level of recovery needed was therefore on a vaster level than envisioned and resulted in restructuring of businesses, cessation of businesses, expansion of businesses and more as previously discussed. Overall, while the adventure tourism community survived and rebounded it was not unscathed. It is also not home free—other disasters have affected the archipelago since 2017 and will continue to do so.

All of the operators interviewed for the study showed a positive outlook for the future of the adventure tourism industry; their positivity is reflected in actions taken to develop new products, capitalize on new opportunities, and forge ahead with expansion plans. The demand for zipline and ropes course experiences and other adventure recreation activities do not appear to be lessening. Each tour operator has made or plans to make changes to capture the continuing interest in adventure recreation activities that interest tourists. Some like *Acampa* are working to anticipate future needs and provide adventure opportunities that other operators do not offer. The restructuring of the industry, the ongoing investments from adventure tourism businesses and the marketing support from Discover Puerto Rico will go a long way in reinforcing the industry's resilience to natural and other types of hazards.

One interesting idea that emerged from the analysis is that the adventure tourism community could benefit from a more formal structure. While their informal network has helped to make the community resilient, vulnerabilities remain which threaten this resilience. Many adventure businesses did not survive Hurricane Maria and banding together would help to provide greater support to the ones that remain. It is therefore recommended that the adventure tour operators create a co-operative or a formal sub-group within the Puerto Rico Hotel and Tourism Association of which a few of them are already members.

This was a small pilot study to explore an embryonic area of research in the Caribbean. Given the potential for growth in adventure tourism across the region and the anticipated increase in frequency and strength of hurricanes impacting the region as a result of climate change, it would be prudent to extend this research across the region. Other facets of this research could evaluate these communities through the lens of Twigg's (2009) community resilience guidance or specific tourism resilience models and evaluate how they could be used to help bolster their resilience. While the emphasis in this chapter was on how adventure tourism businesses were impacted by and responded to Hurricane Maria, it is important to remember that the owners, their families, and friends live in Puerto Rico and were also dealt a tremendous emotional blow from this hurricane. The businesses survived because the people themselves are resilient and while working to keep their businesses afloat, were also working on their own mental and physical recovery. The researchers are in awe of their fortitude and grateful that despite the sometimes painful memories they chose to share their experiences.

Acknowledgements The authors offer sincerest gratitude to our gracious and knowledgeable interviewees:

Anna Mayol del Valle—Foundation Puerto Rico and Discover Puerto Rico

Ian Purcell—EcoQuest Adventures and Tours

Jorge Jorge and Sylvia Rivera—ToroVerde Nature Adventure Park

Luis Acevedo, Ed.D. Association for Challenge Course Technology Certified Inspector

Naireisa Ginés González—Puerto Rico Department of Tourism

Oscar Ramos—Paddle Paradise PR and Outdoor Center

Raymond Sepulveda—Acampa

Funding for this research was generously provided by California State University Northridge's Department of Recreation & Tourism Management, the Center for Recreation & Tourism, and the College of Health & Human Development.

References

Biggs, Duan. 2011. Understanding Resilience in a Vulnerable Industry: The Case of Reef Tourism in Australia. *Ecology and Society* 16 (1). URL: http://www.ecologyandsociety.org/vol16/iss1/art30/.

Biggs, Duan, C. Michael Hall, and Natalie Stoeckl. 2012. The Resilience of Formal and Informal Tourism Enterprises to Disasters: Reef Tourism in Phuket, Thailand. *Journal of Sustainable Tourism* 20 (5): 645–665. https://doi.org/10.1080/09669582.2011.630080.

Caribbean Disaster Emergency Management Agency. 2018. *Building a Caribbean Pathway for Disaster Resilience in the Caribbean Community (CARICOM)*. St. Michael, Barbados: Caribbean Disaster Emergency Management Agency. Accessed December 2020. https://www.eird.org/americas/pr2020/docs/building-a-caribbean-pathway-caricom.pdf.

Caribbean Tourism Organization. 2017. CTO Declares 2017 "Year of Adventure" in the Caribbean News Release, January 10, 2017. Accessed December 2020. https://www.onecaribbean.org/cto-declares-2017-year-adventure-caribbean/.

Casiano Communications Inc. n. d. *QP ¡Qué Pasa! The Official Magazine of the Puerto Rico Tourism Company*. San Juan, Puerto Rico: Casiano Communications Inc., 2013. Various volumes.

Cutter, Susan L., Lindsey Barnes, Melissa Berry, Christopher Burton, Elijah Evans, Eric Tate, and Jennifer Webb. 2008. A Place-based Model for Understanding Community Resilience to Natural Disasters. *Global Environmental Change* 18 (4): 598–606. https://doi.org/10.1016/j.gloenvcha.2008.07.013.

Dahles, Heidi, and Titi Prabawa Susilowati. 2015. Business Resilience in Times of Growth and Crisis. *Annals of Tourism Research* 51: 34–50.

Duffy, Lauren N., Carol Kline, Jason R. Swanson, Mechelle Best, and Hunt McKinnon. 2016. Community Development Through Agroecotourism in Cuba: An Application of the Community Capitals Framework. *Journal of Ecotourism* 1–19. https://doi.org/10.1080/14724049.2016.1218498.

Findlay, Kaia. 2018. Puerto Rico's Natural Revolution after Hurricane Maria. *HuffPost*. Published electronically August 15, 2018. Accessed December 2020. https://www.huffpost.com/entry/puerto-ricos-natural-revolution-after-hurricane-maria_n_5b5f3dffe4b0b15aba9b4daf.

Henderson, Joan, C. 2005. Responding to Natural Disasters: Managing a Hotel in the Aftermath of the Indian Ocean Tsunami. *Tourism and Hospitality Research* 6 (1): 89–96.

Holladay, Patrick J., and Robert B. Powell. 2016. Social-ecological Resilience and Stakeholders: A Qualitative Inquiry into Community-based Tourism in the Commonwealth of Dominica. *Caribbean Studies* 44 (1/2): 3–28.

Holling, C.S. 1973. Resilience and Stability of Ecological Systems. *Annual Review of Ecology & Systematics* 4: 1–23. https://doi.org/10.1146/annurev.es.04.110173.000245.

Jenkins, I. 2019. *Adventure Tourism and Outdoor Activities Management: A 21st Century Toolkit*. Boston, MA: CABI.

Johnson, Sherry. 2015. *The History and Science of Hurricanes in the Greater Caribbean*. In Oxford Research Encyclopedias: Latin American History. Oxford University Press.

Meyer, Robinson. 2017. What's Happening with the Relief Effort in Puerto Rico? *The Atlantic*. Accessed December 2020. https://www.theatlantic.com/science/archive/2017/10/what-happened-in-puerto-rico-a-timeline-of-hurricane-maria/541956/.

Munroe, Thailynn, Nancy Harris, and Tamara Heartsill Scalley. 2018. While Puerto Rico's People Still Suffer the Effects of Hurricane Maria, Its Forests Are Faring Much Better. World Resources Institute. Accessed December 2020. https://www.wri.org/print/64029.

Office for Coastal Management. 2019. Fast Facts: Hurricane Costs. Accessed December 2020. https://coast.noaa.gov/states/fast-facts/weather-disasters.html.

Pasch, Richard J., Andrew B. Penny, and Robbie Berg. 2019. *National Hurricane Center Tropical Cyclone Report: Hurricane Maria (AL152017)*. National Hurricane Center. Accessed December 2020. https://www.nhc.noaa.gov/data/tcr/AL152017_Maria.pdf.

Puerto Rico Tourism Company. 2016. Guinness World Records™ Certifies "El Monstruo" at Toro Verde. News release, March 2. Accessed December 2020. https://www.prtourism.com/dnn/Press-Room/ArtMID/1508/ArticleID/92/Guinness-World-Records%E2%84%A2-certifies-%E2%80%9CEl-Monstruo%E2%80%9D-at-Toro-Verde.

Rindrasih, Erda. 2019. Life after Tsunami: The Transformation of a Post-tsunami and Post-conflict Tourist Destination; the Case of Halal Tourism, Aceh, Indonesia. *International Development Planning Review* 41 (4): 517–540.

Rosenbloom, Stephanie. 2017. The Recovery Process: Caribbean Tourism, Island by Island. *The New York Times*, October 13. Accessed December 2020. https://www.nytimes.com/2017/10/13/travel/caribbean-tourism-hurricane-recovery.html.

Simmons, Ann. 2018. Hurricane Maria Stripped Puerto Rico's Forests Bare. Now Conservationists and Scientists Are Working to Replenish Them. *Los Angeles Times*. Accessed December 2020. https://www.latimes.com/nation/la-na-puerto-rico-environment-20180226-story.html.

Sydnor-Bousso, Sandra, Kathryn Stafford, Michael Tews, and Howard Adler. 2011. Toward a Resilience Model for the Hospitality & Tourism Industry. *Journal of Human Resources in Hospitality & Tourism* 10 (2): 195–217. https://doi.org/10.1080/15332845.2011.536942.

Torres, Ariana P., Maria I. Marshall, and Sandra Sydnor. 2019. Does Social Capital Pay Off? The Case of Small Business Resilience after Hurricane Katrina. *Journal of Contingencies and Crisis Management* 27 (2): 168–181.
Twigg, John. 2009. Characteristics of a Disaster Resilience Community: A Guidance Note, version 2. In DFID Disaster Risk Reduction Interagency Coordination Group. Accessed December 2020. http://discovery.ucl.ac.uk/1346086/1/1346086.pdf.
United Nations Office for Disaster Risk Reduction. 2017. Terminology. Accessed December 2020. https://www.undrr.org/terminology/resilience.
UN World Tourism Organization. 2018. *Compendium of Tourism Statistics Dataset*. Madrid: UN World Tourism Organization. Accessed September 2019. http://statistics.unwto.org/news/2018-03-23/methodological-notes-tourism-statistics-database-2018-edition.
United States Department of Defense. 2017. DoD, Partner Agencies Support Puerto Rico, Virgin Islands Hurricane Relief Efforts. Published electronically September 26. Accessed December 2020. https://www.defense.gov/Newsroom/News/Article/Article/1325245/dod-partner-agencies-support-puerto-rico-virgin-islands-hurricane-relief-efforts/.
World Travel & Tourism Council. 2018a. Caribbean Resilience and Recovery: Minimising the Impact of the 2017 Hurricane Season on the Caribbean's Tourism Sector. Accessed December 2020. https://www.wttc.org/priorities/crisis-preparedness/caribbean-recovery/.
World Travel & Tourism Council. 2018b. Travel & Tourism Economic Impact 2018: Puerto Rico. Accessed December 2020. https://www.wttc.org/economic-impact/country-analysis/country-data/.
World Travel & Tourism Council. 2019. Puerto Rico 2019 Annual Research: Key Highlights. Accessed December 2020. https://www.wttc.org/economic-impact/country-analysis/country-data/.

CHAPTER 8

An Integrated Path Towards a Resilient Tourism Sector in North-East Tobago

Joanna Moses-Wothke, Aljoscha Wothke, and Leslie-Ann Jordan

Introduction

On 28 October 2020, Tobago's North-East region was declared a Biosphere Reserve by the UNESCO Man and the Biosphere (MaB) Programme (UNESCO 2020a). Tobago now joins the World Network of Biosphere Reserves, which currently numbers 714 biosphere reserves in 129 countries around the globe (UNESCO 2020a). Other biosphere reserves include famous destinations like Mount Olympus in Greece and

J. Moses-Wothke
Tobago Hospitality and Tourism Institute, Mount St George, Trinidad and Tobago

A. Wothke
Environmental Research Institute Charlotteville, Charlotteville, Trinidad and Tobago
e-mail: info@eric-tobago.org

L.-A. Jordan (✉)
University of the West Indies, St. Augustine Campus, Trinidad and Tobago
e-mail: leslie-ann.jordan-miller@sta.uwi.edu

A. Lewis-Cameron et al. (eds.), *Managing Crises in Tourism*,
https://doi.org/10.1007/978-3-030-80238-7_8

Komodo National Park in Indonesia (Global Voices 2020). An MaB Reserve is an area that is recognized for its conservation value, cultural heritage, potential for education and research and opportunities for sustainable development. UNESCO's designation is intended to foster a more harmonious relationship between people and nature and this is achieved by reconciling human activity with the conservation of biodiversity and by establishing a scientific basis for the sustainable management of the Biosphere's natural resources (Global Voices 2020). According to UNESCO (2020a), Tobago's North-East region is now the largest MaB site in the English-speaking Caribbean Small Island Developing States (SIDS), the only larger UNESCO Biosphere Reserve in the Caribbean is located in the French territory of Guadeloupe.

North-East Tobago is a rural, underdeveloped and marginalized segment of the 300km^2 island of Tobago. The area is socially and economically vulnerable to external shocks such as climate change, hurricanes, economic crises, the recent COVID-19 pandemic and related fluctuations in the developing community-based tourism sector. At present, the North-East Tobago UNESCO MaB Reserve encompasses 15 villages, from Belle Garden in the South-West to Moriah in the North-West, with approximately 12,000 residents, who are strongly dependent on unemployment relief programmes, governmental employment and unsustainably managed tourism and fisheries (see Fig. 8.1). According to the Tobago House of Assembly (THA), the timing of this designation was beneficial for the destination as it relates to helping to revitalize tourism development and branding it as a premier eco-tourism destination (Loop News 2020). This site encompasses the first-ever Natural National Heritage Site designation for Trinidad and Tobago (designated in October 2018) and the Tobago Main Ridge Forest Reserve which is the oldest, legally protected, tropical rainforest in the world. The UNESCO MaB further includes Important Bird and Biodiversity Areas (IBAs), as well as possesses numerous expressions of art, festivals and significant historical sites. This unique natural and cultural landscape, as well as a sensitive policy change, specifically towards stakeholder participation, are now integrated through the UNESCO MaB Programme in North-East Tobago and it is hoped that this approach can be replicated with other tourism initiatives meant to strengthen the tourism industry's resilience.

This Chapter will analyze the path taken by North-East Tobago between 2015 and 2020 towards the sustainable management of its

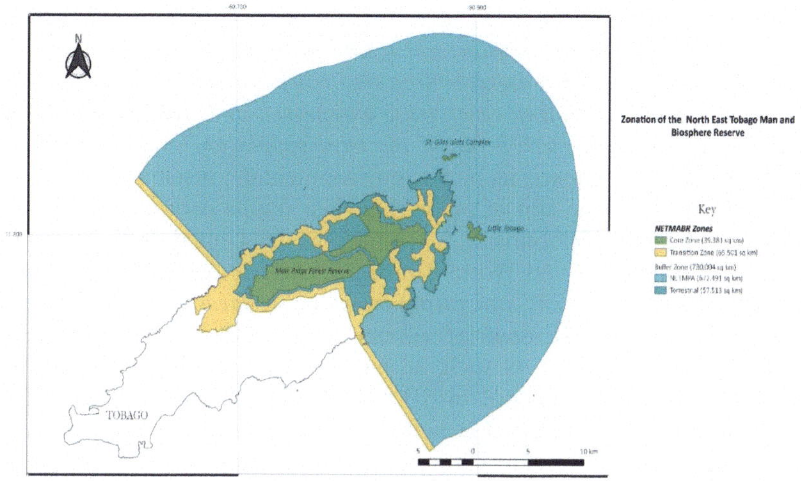

Fig. 8.1 The designated North-East Tobago UNESCO Man and the Biosphere (*Source* Developed by ERIC for the THA, 2019)

natural and cultural resources aimed at supporting cross-sectoral resilience to external shocks while focusing on the critical success factors that allowed Tobago to achieve this designation. More specifically, the Chapter will provide a reflective approach on the process, discussing the role of key stakeholders, institutional hindrances, the legal framework, stakeholder participation and conclude with main lessons learnt. For the purpose of this chapter, Tobago specific information was collected via primary and secondary research; notably, two of the authors were significantly involved as technical experts in the process of meeting important milestones of the journey and finally achieving the UNESCO MaB designation.

Tourism Resilience and Environmental Sustainability

Small Island Developing States (SIDS) like Tobago are faced with significant challenges to their economic development and many have pivoted towards the tourism industry as an engine of growth. However, as noted by Hampton and Jeyacheya (2020, 8), tourism development can be described as a "*'two-edged sword' given that its major economic benefits can also create local vulnerabilities such as an increased demand for water, food, and energy; elevated sewage, waste, and pollution; coastal-zone urbanization and development; overcrowding and traffic congestion; degradation of natural assets, including coral reefs, mangroves, and seagrass meadows; and an erosion of the well-being of the local population*". Consequently, many islands have chosen to pursue environmentally sustainable tourism development with the aim of preserving their fragile natural and marine resources. Sustainable tourism, as defined by the United Nations Environment Programme and World Tourism Organization (2005) "*…takes full account of its current and future economic, social and environmental impacts, addressing the needs of visitors, the industry, the environment and host communities*". As such, sustainable tourism development can help build tourism resilience in SIDS like Tobago and can also offer destinations many additional benefits, including environmental conservation and jobs for women, young people and indigenous communities (Twining-Ward et al. 2017).

A lot of attention has been given to one aspect of sustainable tourism development, that of environmental sustainability. As a heavy resource user (Nelson 2013), tourism is said to have a three-way relationship with the environment: that of conflict, symbiosis and coexistence. However, the literature indicates that conflict between the environment and development initiatives is what led to the discussions on sustainable development (Harris 2003; UNEP 2007; Metz et al. 2007). Today, it remains at the forefront of discussions sometimes overshadowing the other dimensions of sustainability. Although not discounting the importance of economic and socio-cultural sustainability, environmental sustainability seems to be at the core of sustainable development since as stated by Goodland (1995), the maintenance of the environment and its life support system are prerequisites for the other dimensions. So, how can this UNESCO designation help Tobago to develop a more sustainable, resilient tourism industry?

According to UNESCO (2017, 11), the MaB Programme "*…is an important means to mainstream sustainable development at all levels, integrating economic, social and environmental aspects and recognizing their vital interlinkages, in order to achieve sustainable development in all its dimensions*". This designation can also help destinations to "*…strengthen resilience to disaster, including climate change, and protect the natural environment through sustainable use of resources*" (UNESCO iv). What really matters is the interconnectivity among environmental governance, policies and plans using interdisciplinary approaches, the adaptation of tourism businesses to respond to shocks, identification of the value of biodiversity as a tourism asset and the human relationships (Saarinen and Gill 2018), all methodically intertwined, this is responsible for a resilient tourism industry.

Tourism in North-East Tobago

The attractions within the North-East Tobago UNESCO MaB Reserve (NETMABR) are mostly nature-based, ranging from beautiful beaches, ridge to reef scenery, forest and marine biodiversity in combination with the experience of rural fishing-village living. North-East Tobago is the most important terrestrial location for ecotourism in the nation and was voted the "World's Leading Eco-Tourism Destination" by the World Travel Award each year from 2003 to 2006, and the "World's Leading Green Destination" in 2007 and 2009. The Main Ridge Rainforest Reverse, Little Tobago and St Giles Island are an internationally Important Bird and Biodiversity Area and designated as the first National Natural Heritage Site and as a Queens Commonwealth Canopy Site. The Main Ridge Forest Reserve provides aesthetic beauty and recreational attractions which draw local, regional and international tourists and carry equally important heritage and spiritual values. Other attractions, such as cultural festivals and sports events, target the local and national clientele.

The main accommodation providers are small guest houses, privately-owned villas and three small hotels with up to 25 rooms. The village of Castara, for example, is famous for its community-based tourism product and sustainability efforts; the village of Charlotteville is known as an environmental science tourism hotspot; and the village of Speyside is internationally renowned for diving. Unfortunately, there is no data regarding visitation numbers specifically for North-East Tobago. The main, domestic tourism market for Tobago is Trinidad; however, the vast majority of domestic tourists are staying on the south-western side of the island which is more geared towards entertainment and nightlife-based tourism. Local tourists visiting the proposed NETMABR area are single-day visitors on a road trip through the island and spending money on mainly catering.

The annual cruise ship arrivals in Tobago fluctuate significantly (e.g. in 2017–2018 at 82,000; in 2016–2017 at 23,821). Tours to the Main Ridge Rainforest Reserve and Little Tobago are offered and booked but it can be argued that the economic, social and environmental impact of cruise ship tourism in North-East Tobago is relatively low. The number of overnight stays, staying within North-East Tobago can only be estimated at a rough order of magnitude scale. The room stock in North-East Tobago is estimated at approximately 400 rooms with an average booking of 40% resulting in roughly 160 daily overnight stays. The average cost per room is 60 USD, resulting in income from an accommodation of roughly 9,600 USD/day for the entire NETMABR area. Further income, by guess of similar value, is generated via car rentals, taxi services, catering, supermarket spending, guided tours and diving.

Despite the generally low visitation level for entire Tobago, the efforts to establish sustainable community-based and science tourism products in North-East Tobago have led to a relatively significant increase in bookings over the past five years, especially in Castara and Charlotteville. Due to the perceived relatively low number of visitors, the negative impacts of tourism are currently limited. Neither reefs nor forests are visited beyond the limits of acceptable change. The pressure on natural resources such as water, fish, fruits, vegetables, and other commodities is also limited given that the average visitor number is less than 3% of the total population at one given time. The same applies to the very limited increase of waste produced by overnight visitors, which are often environmentally conscious. Spearfishing tourism (mainly by Trinidadian divers) has led to a decrease in mesopredators at specific reefs; however, restrictions on recreational spearfishing are envisaged for future natural resource management plans.

Case Study Area—North-East Tobago

The North-East Tobago UNESCO Biosphere Reserve presents a rare, largely intact Caribbean island ridge-to-ocean ecosystem that includes the Tobago Main Ridge Forest Reserve, which is the world's oldest, legally protected tropical rainforest reserve, established in 1776. It encompasses 83,488 hectares, with a marine area of 68,384 ha that is home to coral reefs and mangroves. Overall, 1,774 species have been recorded in its 19 habitat types and it is home to globally unique and endangered plants and animals including 83 International Union for Conservation of Nature (IUCN) Red List of Threatened species and 41 endemic species (Global Voices 2020; UNESCO 2020a). The Biosphere's 19 distinct habitat types are home to globally unique and endangered species. While the avifauna is highly diverse, the highest faunal endemism is found among the herpetofauna which includes three snakes, one lizard and four frogs. The Reserve is also home to one of the last remnants of a dry tropical forest in the Caribbean. About 12,000 people live in this area. The inhabitants share a culturally, deeply rooted socio-economic and spiritual relationship with the natural resources. Through folk stories, traditional medicines, and ceremonies the importance of this land can be shown. The 15 villages in the area are economically dependent on artisanal fishing, tourism, forestry and governmental employment (around 60%) in the administrative, public service or unemployment relief sectors (UNESCO 2020b).

The eligibility for the UNESCO MaB was also supported by North-East Tobago's rich colonial and slavery-related history and the resulting cultural and socio-economic activities. The terrestrial part of the planned NETMABR has historically been used extensively for sugar, coconut and cocoa-based agriculture, which is indicated by many colonial ruins and, in the case of cocoa, still within living memory. The need for an ecosystem service namely consistent and sufficient water supply for the sugar industry was the main reason for the declaration of the Main Ridge Forest Reserve in 1776. The cultivation of sugar, coconut, cotton, and cocoa for export formed part of the agricultural history, a common post-colonial socio-cultural and economical history typically found in the southern Antilles SIDS. The communities within the NETMABR were developed out of slave and colonist settlements within several large estates. Today, land use is still dominated by abandoned agro forestry estates, while the ocean was, and still is increasingly used for artisanal fishing. The breakdown of the agricultural estates during the last century and the dependence on agricultural imports in the past decades are two main factors for the transformation in the economic landscape. Tobago's political structure is still influenced by the former way of colonial administration and has historically neglected the importance of civil society inclusion. The micro and small business sector does not have strong representation through associations or a business chamber. However, the number of persons involved in agriculture and fisheries tend to gain political attention during election periods.

The Journey to the UNESCO Designation

Between 2013 and 2020, a series of mutuality reinforcing initiatives unfolded in North-East Tobago targeting sustainable development, natural resource management and stakeholder participation. The combination of these initiatives (partly coincidental) was historically unprecedented and laid the foundation to develop more detailed management plans focusing on building resilience to external shocks and creating the basis for responsible tourism product development, branding and marketing as part of economic development. At the base of cross-sectoral resilience to external shocks is the health of the resident ecosystem services provided to and used by all sectors of society for the following:

- provision of services such as food, construction material, water;
- regulation of services such as water control, pest control, climate regulation;
- cultural services such as recreation, relaxation, education, research; and
- support services such as biodiversity maintenance and ecosystem health.

Only if these ecosystem services are functional, all sectors of society can achieve their goals and as such all sectors are important stakeholders in the management of these services. In fragile environments and vulnerable communities in Caribbean SIDS, such as (North-East) Tobago, the role of tourism is significant. In the North-East Tobago case, tourism has been on par with fishing as the top two private sectors, economic contributors (pre-COVID). During the COVID-19 pandemic, fishing kept many fishing villages economically afloat and tourism has been earmarked as one of the most potent economic recovery tools post-COVID.

Several key initiatives that occurred between 2013 and 2020 laid the foundation to support North-East Tobago's ecosystem services and ultimately lead to the UNESCO designation. These activities provided the ingredients for sustainable blue, green and purple economic development especially in the responsible tourism sector including, but not limited to community-based tourism, small boutique resorts, dive tourism, ecotourism, science/educational tourism, cultural/historical tourism and health tourism. These key initiatives were the Improved Forest and Protected Area Management project (2013–2020), the establishment of the North-East Tobago Protected Area Management Trust (2019) and the designation of the Main Ridge Forest Reserve as the nation's first National Natural Heritage Site.

Improved Forest and Protected Area Management Project

In line with the adoption of the new protected area and forest policies in 2011 and a New National Wildlife Policy in 2013, the Government of Trinidad and Tobago (GORTT) received financial support from the Global Environmental Facility (GEF) through the Food and Agricultural Organization of the United Nations (FAO) to implement the Improved Forest and Protected Area Management (IFPAM) project. Tobago's Main Ridge Forest Reserve and a planned North-East Marine Protected Area were the target areas in Tobago under this national project. The project was implemented between 2013 and 2020 and, inter alia, laid many foundations for North-East Tobago's path towards sustainable development, natural resource management, resilient and sustainable tourism and stakeholder involvement. Specifically, management plans for the Main Ridge Forest Reserve and the planned MPA were developed, a new National Protected Area Systems Plan (including many new sites in North-East Tobago) consolidated and accepted by Cabinet, knowledge, attitude and practice (KAP) analyses conducted and capacity built for stakeholders.

One of the most significant impacts was the establishment of a Tobago Project Steering Committee, composed of representatives of relevant THA Departments as well as civil society organizations (CSOs). Despite serious design flaws, especially regarding the meaningful sharing of intelligible information and CSO compensation, this Steering Committee was the longest collaboration between state and non-state actors regarding conservation and sustainable natural resource management in the history of Tobago. As such, they were able to create relationships and networks that have survived the IFPAM project and are still able to influence development decisions. A further, significant by-product, initiated by IFPAM stakeholders, was the establishment of the North-East Tobago Protected Area Management Trust, envisaged to manage protected areas and implement the UNESCO MaB programme in North-East Tobago.

North-East Tobago Protected Area Management Trust

Since 2017, North-East Tobago stakeholders, especially those who participated in the IFPAM project, agreed that a participatory management approach and facilitating the development of sustainable blue and green economies are the appropriate answers to North-East Tobago's challenges with positive spill-over effects on the entire island and country. As a result, as well as officially mandated by an Executive Council Note of the THA, an interim, multi-stakeholder Board established the North-East Tobago Protected Area Management Trust (NETPAMT) on 11 February 2019. The THA committed to vest the management authority for nationally protected areas in North-East Tobago with the NETPAMT as well as entrust it with the management of the UNESCO MaB Programme. The NETPAMT is envisaged as a co-management organization fully managing legally protected areas in North-East Tobago as well as implementing the North-East Tobago UNESCO Man and the Biosphere Reserve programme through facilitating cooperation with state agencies, civil society and the private sector.

This nationally unprecedented institutional transition from a post-colonial natural resource management approach to a participatory, transparent, accountable co-management model is demonstrably international best practice and fully in line with government policies. Inter alia, the role of NETPAMT is described as follows:

i. Assume responsibility for Protected Area Management Plans;
ii. Implement projects and programmes that support the integrity, viability, sustainability of the ecosystems vested in the Trust;
iii. Secure other funding sources outside of THA subventions and collectable fees;
iv. Promote public awareness of the ecological systems and natural resources of North-East Tobago Protected Areas and their importance in sustaining livelihoods on the island;
v. Promote and support sustainable livelihoods of community members in and around North-East Tobago Protected Areas through aggressively pursuing synergies with established initiatives and utilizing innovative approaches;
vi. Drive the development and promotion of North East Tobago Protected Areas as lead ecotourism products.

While the framework and mandate for the NETPAMT have been described, the organization became dysfunctional in late 2019. However, the current designation of North-East Tobago as a UNESCO Man and the Biosphere Reserve has brought the necessity to revitalize the Trust to the forefront and various stakeholders are currently working on developing proper governance structures with multilateral donor assistance.

Designation of the Main Ridge Forest Reserve as the First Natural National Heritage Site of Trinidad and Tobago

In 2017, the National Trust of Trinidad and Tobago listed Tobago's Main Ridge Forest Reserve as a National Heritage Asset and approached the THA to conclude the designation process by preparing a dossier outlining the justification why the Main Ridge Forest Reserve should be declared a National Heritage Site and to be presented to Cabinet. The Division of Infrastructure, Quarries and the Environment (DIQE) and the National Trust approached the Environmental Research Institute Charlotteville (ERIC, http://www.eric-tobago.org), which was working on the UNESCO MaB nomination form in early 2019, to assist with the preparation of the required dossier. The dossier covered relevant aspects such as biogeography and ecology, biodiversity, connectivity, hydrology, geology, climate, historical background, folk tales, ecosystem services, threat, conservation barriers, stakeholders, legal and regulatory aspects and recommendations for sustainable management. In May 2019, the ERIC submitted the dossier to the THA which forwarded it to the National Trust of Trinidad and Tobago. The process was finalized by the official listing of Tobago's Main Ridge Forest Reserve as a "Property of Interest" in the Trinidad and Tobago Gazette on 5 December 2019. Remarkably, this designation and the improved conservation status of the Main Ridge Forest Reserve under the National Trust Act, was achieved in less than eight months! As a result, the Main Ridge Forest Reserve now enjoys a higher level of protection and an additional branding and marketing tool was provided to, mainly community-based, tour guiding operators in North-East Tobago.

UNESCO MaB Designation

In August 2011, the Trinidad and Tobago National Commission for UNESCO submitted a tentative listing of the Tobago Main Ridge Forest Reserve, the world's oldest, legally protected tropical forest, for consideration as a UNESCO World Heritage Site (http://whc.unesco.org/en/tentativelists/5646/). The process of submitting a full application never started due to a lack of commitment of state agencies on the national as well as on the Tobago level; furthermore, the justification for listing was extremely weak. Over the years, the national Tourism Development Company (TDC) had taken an interest in restarting the World Heritage Site designation process; however, in 2017, when the TDC was dissolved, the designation process lost its final formal champion. On its final days, key staff members of the TDC asked the Environmental Research Institute Charlotteville, to pick up the trail and re-engage with stakeholders.

Fortunately, this happened under the following enabling circumstances. The previously mentioned Improved Forest and Protected Area Management (IFPAM, https://www.protectedareastt.org.tt) project, which focused on North-East Tobago and specifically on the Main Ridge Forest Reserve, had created heightened political and broad stakeholder interest in the area. Under a new administration, the THA integrated the chronically under-resourced Department of the Environment into the powerful and financially more liquid Division of Infrastructure, Quarries and the Environment (DIQE) under the leadership of a Secretary with a strong interest in sustainability, conservation and responsible tourism. The Director of the DIQE was willing to put his trust into the technical expertise of a relatively young environmental NGO and collaborated with the ERIC in approaching the Trinidad and Tobago National Commission for UNESCO to explore the possibilities for a new start in the designation process.

In January 2018, the DIQE signed a contract, securing technical expertise from the ERIC to assist with the preparation of documents for submission to the UNESCO World Heritage Committee. The first steps involved: stakeholder registration, stakeholder engagement plan and communication plan, followed by an evaluation to guide the THA and other key stakeholders regarding the selection of the most appropriate, desirable, acceptable and achievable form of UNESCO recognition for North-East Tobago's Natural Heritage, simultaneously benefitting conservation and sustainable development. This evaluation included a site description, threats and opportunities, a regional gap and comparative analysis, key stakeholder interviews and surveys and an evaluation whether a Man and the Biosphere (MaB) or a World Heritage Site status would be more appropriate for North-East Tobago.

The evaluation provided much-needed clarity, determining that a MaB status would be realistically achievable and much better aligned with stakeholders' interest in strategic sustainable development in conjunction with conservation and utilization of natural and cultural heritage. Furthermore, it became clear that a potential UNESCO site should include the entire north-eastern end of Tobago including a marine area with a 11.1 km seawards boundary (which is the area that falls under the management authority of the THA). At the same time, the regional gap and comparative analysis referring to other UNESCO World Heritage Sites as well as key expert interviews indicated that there was no scientific evidence that North-East Tobago possessed the global significance and highest level of Outstanding Universal Values to meet the criteria required by the UNESCO World Heritage Convention. Based on the above findings, the focus was put on completing the UNESCO MaB Nomination form before its deadline in September 2019.

The nomination document included a detailed site description and analysis and explored the alignment of development policies and stakeholders' interests with the UNESCO MaB Programme's interconnected functions of conservation, development and logistics (education, research and knowledge exchange). All relevant Departments of the THA were interviewed and stakeholder consultations were conducted with CSOs during open community meetings (though not very well attended, due to very limited implementation of the related communication plan). Most importantly, the application included the THA approved zonation of North-East Tobago into three MaB management zones: the Core Zone, comprising of already legally protected areas; the Buffer Zone, surrounding the core areas and earmarked for low sustainable development; and the Transition Zone, including a 125 m wide band alongside all roads and settlements.

Important spin-off effects of this process were the declaration of Tobago's Main Ridge Forest Reserve as the nation's first Natural National Heritage Site and the first complete register of all documented species in North-East Tobago. In September 2019, the nomination document was approved by the THA and submitted by the Ministry of Planning via the Trinidad and Tobago Commission to UNESCO, the Ministry of Foreign and CARICOM Affairs and the Trinidad and Tobago Mission to UNESCO to the UNESCO MaB Headquarters in Paris. Delayed due to the COVID-19 pandemic, North-East Tobago was finally declared a UNESCO Man and the Biosphere Reserve on 28 October 2020 and the Government of Trinidad and Tobago was congratulated for a well-prepared application.

Discussion and Conclusion

The main benefit of the UNESCO MaB designation and the implementation of its programme in North-East Tobago is the integration of conservation, sustainable development (including tourism), research and education as a superstructure to existing management practices and policies. Within this context, sustainable green, blue and purple economic activities in North-East Tobago, including but not limited to tourism, fisheries, agriculture, cultural heritage activities, research and education will be incentivized (Loop News 2020). Environmentalists in T&T are hopeful that the title of UNESCO Biosphere Reserve will incentivize locals to diversify the economy towards sustainability in order to retain the designation for future generations (Global Voices 2020). The UNESCO MaB Programme is a perfect tool for economic recovery after the COVID-19 pandemic especially for tourism and attracting international support (Loop News 2020). Secretary for the DIQE, Councillor Kwesi Des Vignes, stated that "*The award could not come at a better time for Trinidad and Tobago…As the world grapples with COVID- 19, this designation has placed us in an advantageous position as it relates to [the] branding of our destination and attracting financial and technical support from the international community. We in Tobago in particular, are excited and proud of this accomplishment and the many gains that will follow.*" (Trinidad and Tobago Guardian 2020a). He also stated that "*This award presents immense opportunities for business, culture and the arts and education, but more importantly, it will allow us to benefit from approaches that can better conserve and sustain the environment and experiences we gain from interacting with it*" (Trinidad and Tobago Guardian 2020b).

Once well implemented, the UNESCO Site will be Tobago's strategic lead investment to prepare an enabling institutional and organizational environment for future, scaled-up investment by multilateral agencies and the private sector in sustainable, ecosystem-based blue and green economy in North-East Tobago, with significant beneficial spill-over effects on Tobago as a whole and the entire nation (Loop News 2020). According to Minister Robinson-Regis, "*We are continuing to punch above our weight in the global arena. We are blessed with some of the richest historical, cultural and natural heritage in the world, and our pride and sensibility in protecting our gifts will continue to be rewarded.*" The Minister also added that protecting Tobago's biodiversity resources also increases the destination's foreign exchange earning potential, because the 5th National Report of Trinidad and Tobago to the Convention on Biological Diversity (CBD) in 2017 revealed that T&T's resources are valued in the region of US $100 million per year in terms of soil protection, water purification services as well as recreation and tourism-based activities (GORTT 2020).

Ancil Dennis, Chief Secretary in the THA (2020) also welcomed the news of the award, calling it proof that sustainable development for Tobago is paramount: "*This Man and the Biosphere designation is only an opportunity for us to improve the management and protection of our natural and cultural heritage — two areas that have always been extremely important to us here in Tobago.*" In another THA video celebrating the milestone, spearfisher Ojani Walker said Tobago's UNESCO designation was a great incentive for the recovery of beaches, reefs, and the environment, while certified tour guide Darlington Chance believed it would bring a lot of interest to Tobago and help generate income for the island's tourism sector. "*If we don't protect [nature], we will not have it*," he added, as he laid out his vision for a sustainable future. Similarly, Patricia Turpin, former President of the NGO, Environment Tobago, having lived in North-East Tobago for the last 45 years, called on fellow residents to "*work together with those who are running this programme*" to conserve the beauty and biodiversity of the area, even as the island grapples with the effects of the climate crisis. As far as environmentalists are concerned, the UNESCO designation should inspire both an urgency to conserve and a sense of national pride for every citizen of Trinidad and Tobago.

One of the most obvious positive impacts of the anticipated increase in tourism is an increase in direct and indirect income for NETMABR residents and Tobago as a whole. Additionally, and once properly managed, increased science and educational tourism can lead to enriched scientific research, knowledge exchange, increase scholarships and publications relevant to sustainable development and management of cultural and natural resources. Within this context, this chapter, in conjunction with related publications and dossiers, is meant to provide the context for further research on the progress of the MaB Programme in Tobago. Scientists now have the opportunity to connect with the monitoring and evaluation processes for the Reserve, link with local and national research institutions and contribute meaningfully to sustainable development and tourism resilience through communicating applicable research results to local stakeholders.

Since 2015, tourists have already been paying for participation in ecosystem health monitoring through programmes such as Reef Check and Forest Check, consequently contributing to conservation efforts. The more residents understand that their income directly depends on a healthy and aesthetically pleasing environment, the more likely they are to protect the area's cultural and natural heritage attractions and assets. Also, agro-tourism activities surrounding cocoa production, processing, chocolate making and export associated (which has already been successfully demonstrated within the area) will be encouraged as it can significantly contribute to sustainable development; and existing governmental research and training facilities now have the potential to be used for knowledge sharing and as role models for other SIDS. With a UNESCO MaB designation, the expansion of responsible tourism as a viable and appropriate economic activity for North-East Tobago is paramount. Once properly managed and marketed, the UNESCO MaB brand will assist in increasing income from tourism to North-East Tobago. However, this will, inter-alia, require the upgrading of visitor facilities, trails, improved waste management systems, customer service training and certifications according to international standards.

There are specific key lessons regarding resilient and sustainable tourism that can be learnt from Tobago's journey to a UNESCO MaB designation. The development of a resilient and competitive sustainable tourism sector in a Caribbean SIDS requires political will, an enabling environment, time and patience, broad stakeholder consensus and strong champions. Figure 8.2 highlights some of the special considerations, in the case of North-East Tobago, that created an enabling environment to support ongoing sustainable development efforts, related to protected areas and facilitated by a broad range of state and civil society actors. While the idea of a UNESCO designation lingered for almost a decade, only the ability of key stakeholders to recognize a rare window of opportunity provided in 2018, allowed them to successfully collaborate and enter the UNESCO designation process accompanied by many other supporting initiatives. The increased public-private sector collaboration between the State and CSO's actors between 2014 and 2020 created the level of trust to achieve consensus on the most appropriate path towards sustainable development, resilient and responsible tourism and conservation of ecosystem services. Most importantly, this path was cleared by a handful of highly dedicated, competent and recognized state and CSO's representatives that chose to collaborate at the right moment.

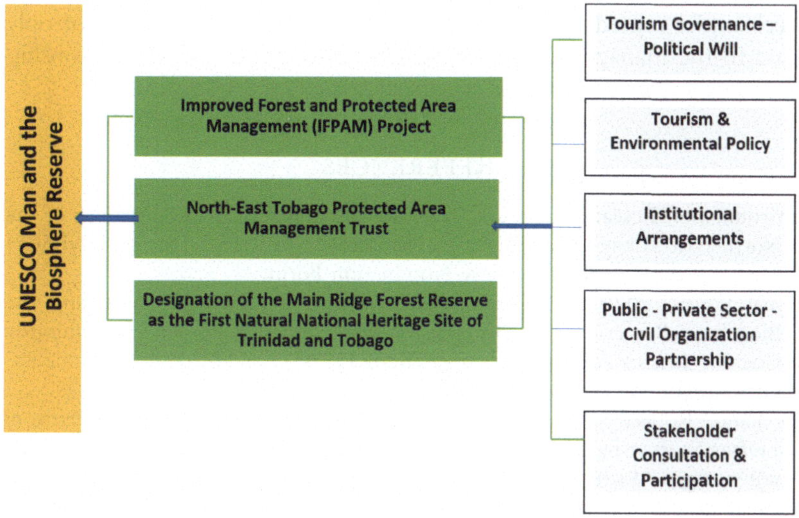

Fig. 8.2 Special Considerations for the UNESCO MaB Reserve Designation (*Source* Authors, 2021)

Institutional arrangements for tourism in SIDS that strongly influence inter-organizational co-operation and sharing of power and resources (Jordan 2007) also helped to support and fast-track sustainable tourism initiatives in Tobago's case. This new tourism landscape described along with anticipated positive and negative impacts, should be proactively managed through the establishment of regulations and setting limits of acceptable change. As was evidenced in this case study, the presence of supportive tourism-related and environmental policies and institutions was critical to building a foundation that supported Tobago's mission of fostering environmentally sustainable tourism. Finally, a participatory management approach, with adequate stakeholder consultations is needed for future progress and success. The future management organization of the NETMABR should have the mandate to coordinate its own (Biosphere focused) management activities with those of the THA Divisions responsible for the various aspects of reduction and mitigation of impact risks. Therefore, it is of utmost importance that the relationship between the NETMABR management organizations and other management authorities is clearly defined and does not allow room for uncertainties regarding responsibilities and mandates; this does not only relate to the management of impacts but also to the overall relationship between these organizations.

References

Environmental Research Institute Charlotteville (ERIC). 2019. Zonation of the North-East Tobago MaB Reserve Document: North-East Tobago UNESCO Man and the Biosphere Reserve Nomination Form.

Global Voices. 2020. Northeast Tobago Named a UNESCO 'Man and the Biosphere Reserve': The Aim Is for Man and Nature to Harmoniously Coexist. https://globalvoices.org/2020/11/05/northeast-tobago-named-a-unesco-man-and-the-biosphere-reserve/.

Goodland, Robert. 1995. The Concept of Sustainability. *Annual Review of Ecology Systematics* 26: 1–24. Accessed April 22, 2019. https://sustainability.water.ca.gov/documents/18/3407876/The+concept+of+environmental+sustainability.pdf.

GORTT. 2020. Tobago Officially Awarded Coveted International UNESCO Designation. http://www.news.gov.tt/content/tobago-officially-awarded-coveted-international-unesco-designation#.X9FxhbmSlyw.

Hampton, M.P., and Jeyacheya, J. 2020. Tourism-Dependent Small Islands, Inclusive Growth, and the Blue Economy. *One Earth* 2, January 24, pp. 8–10, Elsevier Inc. https://www.cell.com/one-earth/pdf/S2590-3322(19)30276-3.pdf.

Harris, Jonathan. 2003. Sustainability and Sustainable Development. Accessed June 5, 2019. http://isecoeco.org/pdf/susdev.pdf.

Jordan, L. 2007. Interorganisational Relationships in Small Twin-Island Developing States in the Caribbean—The Role of the Internal Core-Periphery Model: The Case of Trinidad and Tobago. *Current Issues in Tourism* 10 (1): 1–32.

Loop News. 2020. It's Official: North-east Tobago DECLARED UNESCO Biosphere Reserve. https://www.looptt.com/content/its-official-north-east-tobago-declared-unesco-site.

Metz, B., O.R. Davidson, P.R. Bosch, R. Dave, and L.A. Meyer. 2007. *Contribution of Working Group III to the Fourth Assessment Report of the Intergovernmental Panel on Climate Change*. Cambridge: Cambridge University Press.

Ministry of Tourism. 2020. Revised National Tourism Policy (Draft), July 2020. http://www.tourism.gov.tt/Portals/0/Documents/Press%20Releases/Revised%20Draft%20National%20Tourism%20Policy%20(2020-2030).pdf?ver=q3x37qPEOiT9ReRaCEsgBg==.

Nelson, Velvet. 2013. *An Introduction to the Geography of Tourism*. Plymouth: Rowan and Littefield Publishers.

Saarinen, J., and A.M. Gill. 2018. *Resilient Destinations and Tourism: Governance Strategies in the Transition towards Sustainability in Tourism, Contemporary Geographies of Leisure, Tourism and Mobility*. London: Routledge.

Trinidad and Tobago Guardian. 2020a. North East Tobago Gets UNESCO Award, October 28, 2020. https://www.guardian.co.tt/news/north-east-tobago-gets-unesco-award-6.2.1240838.fb72ed7341.

Trinidad and Tobago Guardian. 2020b. UNESCO Award Gives Tobago Tourism Boost, October 30, 2020. https://www.guardian.co.tt/news/unesco-award-gives-tobago-tourism-boost-6.2.1241682.8da0d53210.

Trinidad and Tobago Newsday. 2020. Tobago Acquires UNESCO Man on the Biosphere Status. https://newsday.co.tt/2020/10/28/tobago-acquires-unesco-man-on-the-biosphere-status/.

Twining-Ward, L., J. Perrottet, and C. Niang. 2017. Resilience, Sustainability, and Inclusive Growth for Tourism in the Caribbean. https://blogs.worldbank.org/psd/resilience-sustainability-and-inclusive-growth-tourism-caribbean.

United Nations Environment Programme (UNEP). 2007. Chapter 1: Environment for Development. Accessed June 5, 2019. http://www.unep.org/geo/geo4/report/01_Environment_for_Development.pdf.

United Nations Environment Programme and World Tourism Organization. 2005. *Making Tourism more Sustainable. A Guide for Policy Makers*. Paris and Madrid: UNEP and UNWTO.

UNESCO. 2017. *A New Roadmap for the Man and the Biosphere (MAB) Programme and Its World Network of Biosphere Reserves: MAB Strategy (2015–2025)*, Lima Action Plan (2016–2025), Lima Declaration. https://unesdoc.unesco.org/ark:/48223/pf0000247418.

UNESCO. 2020a. North-East Tobago Declared as UNESCO Biosphere Reserve, October 28, 2020. https://en.unesco.org/news/north-east-tobago-declared-unesco-biosphere-reserve.

UNESCO. 2020b. North-East Tobago Biosphere Reserve, Trinidad and Tobago. https://en.unesco.org/biosphere/lac/northeast-tobago.

CHAPTER 9

Grenada: A Vision of Integrated Technological Advancements to Build a Resilient Tourism Future Through Youth Involvement and Consumer-Centric Service Excellence

Kimberly Thomas-Francois, Ye Shen, and Marion Joppe

Introduction

Grenada, popularly known as the Isle of Spice of the Caribbean, remains one of the most resilient tourism-dependent destinations of the south-eastern Caribbean. Consistent with neighboring islands, the predominantly agrarian economy has shifted dramatically to a service economy mostly driven by tourism over the past decades (Clayton

K. Thomas-Francois (✉) · M. Joppe
University of Guelph, Guelph, ON, Canada
e-mail: kthoma08@uoguelph.ca

M. Joppe
e-mail: mjoppe@uoguelph.ca

Y. Shen
California Polytechnic State University, San Luis Obispo, CA, USA
e-mail: yeshen@calpoly.edu

A. Lewis-Cameron et al. (eds.), *Managing Crises in Tourism*,
https://doi.org/10.1007/978-3-030-80238-7_9

and Karagiannis 2008; Rhiney 2016). Over recent decades, the island's economy succumbed battering from several economic shocks due to world economic crises such as that of 2001 and 2008, together with devastation from consecutive monstrous Hurricanes Ivan and Emily in 2004 and 2005, respectively. The first resulted in the decimation of almost 90% of the island's housing stock among other destructions to all economic sectors, including significant losses in tourism (Antoine et al. 2014). The destination's economy managed to bounce back each time to a certain extent despite years of volatility (Antoine et al. 2014). Pre-pandemic, the country struggled with a high unemployment rate of 21.7% as recorded in 2017, with youth unemployment being almost twice as high. More recently, signs of a downward trend in unemployment were recorded (National Plan Secretariat 2019).

Tourism in Grenada was positioned for increased growth and further expansion of its international market share until the COVID-19 pandemic shock. This positioning included renovation of older properties and the establishment of new internationally renowned luxury resort brands (CTO 2020) through foreign direct investments. The Caribbean Tourism Organization (CTO) recognized the destination's stellar tourism performance in 2019, as total visitor arrivals broke the record of previous years with over 0.5 million visitors of which 31% were stayover tourists, accounting for an increase in visitor spending of 1% and representing a 10% increase in arrivals in 2018 (CTO 2020). In 2019, there was an increase from 186,000 to 188,000 in overnight stays, but a decrease from 343 to 338 K in same day visits, for a slight total decline from 529 to 526 K (Statista 2020a). Surprisingly, revenues declined much more significantly from USD$ 541.6 million to USD$ 489.6 million in 2019 (Statista 2020b). Prospects for inter-sectorial linkages and positive economic impacts were promising and poised to benefit the construction sector, youth employment and all other tourism-dependent support services in the value chain.

Based on the susceptibility of the tourism sector to external and internal shocks, it is evident that built-in resilience that integrates and spreads dependence and resources for sustainable development in all sectors of the economy is now an imperative. The lessons are recurring, and so the destination's strategy ought to change to ensure diversification in a manner that utilizes human capital to mitigate current challenges. COVID-19 has forced the world to adapt to contingencies using available technologies and creating new ones. The call to re-invent, revolutionize and transform Caribbean tourism remains (McLeod et al. 2017) but goes beyond physical tourism infrastructure. Countries such as China, although no match to Grenada in scale or economy, have shown how policy intervention and support to encourage businesses to integrate internet and other technologies contribute to dynamic internationalized businesses and opportunities (Greeven and Wei 2017).

In this chapter we seek to put forward strategies that complement current tourism products and services and at the same time identify opportunities for creative industries that can ensure sustainable continuity of life in times of disruptions. These strategies are entwined in human capabilities through the use of the internet and other technological advancements. These will be targeted at four levels: tourism product and service enhancements or substitutes; the development of business ecosystems that also support tourism; targeting of new and existing markets; and consumer service customization. More specifically, these strategies will draw on existing technology capabilities from different sectors worldwide such as virtual retailing that involves access and consumption of tourism services, food and cultural products through business ecosystems, the introduction of technologies such as virtual reality for tours and attraction sites, digital identification of tourism consumers to support customization of services, and new modes for accessing destination marketing intelligence. A combined research design is employed using content analysis to review experiences and developments worldwide while also drawing on relevant information and multidisciplinary published cases (Tight 2017). The chapter should therefore serve as a point of reference for Grenada and other regional destinations post-pandemic and beyond.

Literature Review: Re-imaging Tourism and Tourism Resilience

In turbulent times, the exploration of existing facts, retrofitting ideas based on new knowledge, experience, and information, and identifying new ways of problem solving are relevant strategies to tackle the challenges at hand. Understanding the research context is also essential especially in cases with applied research. How then might tourism be re-imagined? And what are the opportunities for resilience in small developing states? This literature mostly emanates from relevant scholarly contributions in two special issue Journals published before and during the COVID-19 pandemic. It emphasizes scholars' recommendation for a re-imaged and a resilient tourism sector beyond the pandemic.

Sheller's (2020) adamant call for the Caribbean to rebuild more sustainable ecologies and economies as the islands recover from the effects of the pandemic while still facing an ongoing climate crisis was the major theme for re-imagination and resilience. There is general agreement that the tourism industry needs to take advantage of the current hiatus in visitor arrivals and build back more sustainably. According to Benjamin et al. (2020), it is an opportunity to redesign the tourism landscape to one that is more sustainable and equitable: a reset that is not just green-washing but planning for a tourism future guided by an ethics of care, social and environmental justice, and racial reconciliation (Benjamin et al. 2020, 479). The emergence of innovative alternative social systems that promote values as prescribed by Benjamin et al. (2020) was also postulated by Lew et al. (2020) in their vision of travel and tourism post pandemic, although stated in terms of equity, fairness and cooperation geared toward protecting vulnerable groups.

Theorists also did not forfeit the opportunity to conceptualize workable frameworks. For instance, in revisiting the concept of resilience, the resilience adaptive cycle was organically described as:

- Phase 1: (Re-) Organization (Innovation and Creativity);
- Phase 2: Growth (exploiting opportunities that arise from Phase 1);
- Phase 3: Consolidation (establishing fixed institutions and rules for Phase 2);
- Phase 4: Collapse (future of fixed institutions in Phase 3 to adapt context changes) which results in a return to the re-organization in Phase 1 or alternatively the complete dissolution of the system (Lew et al. 2020, 456).

Likewise, in re-imaging tourism there were some practical cases that are adaptable to Caribbean tourism. Such cases are built around the collaborative economy advocated by Botsman and Rogers (2010) and Sundararajan (2016). The collaborative economy is seen as an equitable alternative to other capitalist models, and one that is tolerant to open access and improvement to the common (Rifkin 2014). This was exemplified in the case presented by Gyimóthy and Meged (2018) of the Camono, a walking trail communitarian initiative in Denmark that promoted a business model that is alternative, bottom-up and succeeds within principles of the sharing economy. In re-imaging tourism, a common thread in the literature is development alternatives and diverse economies that promote and contribute to reduced vulnerability of destination populations, build local capabilities, create employment for locals and increase opportunities for small businesses (Amoamo et al. 2018; Cave and Dredge 2018; Hughes and Scheyvens 2018). A thorough distinction of Corporate Social Responsibility (CSR) and its link to diverse economies is established. CSR in the tourism sector is often seen as a company's responsiveness to the triple bottom line of financial, social and environmental outcomes (Hughes and Scheyvens 2018, 516). However, the role of beneficiaries of tourism in this context is critiqued as passive (Ashley and Haysom 2006) since communities are less engaged in influencing the outcomes (Hughes and Scheyvens 2018). The essence of diverse economies is considered to be a broader concept than CSR since they represent the diverse interests, values and outcomes of various stakeholders (Gibson and Graham 2008). Therefore, the idea of revisioning and re-imaging tourism according to the alternative and diverse economy perspectives goes beyond the miniscule CSR activities that big businesses in the global tourism value chain engage in. Instead, there should be efforts in destinations that contribute to more meaningful empowerment of the respective tourism-dependent destinations and communities.

The introduction and growth in digital technology have been noted as a medium that can harness economic and social transformation, thereby creating opportunities for new economic spaces that are unconventional (Cave and Dredge 2018; Gibson-Graham 1996, 2006). Technology and technological innovation is expected to emerge as a means of managing COVID-19 and future diseases, and to help in achieving a travel and tourism industry that is more resilient (Zeng et al. 2020). There are, however, some reservations to technology since it reduces human touchpoints. In some countries, artificial intelligence (AI) and other similar technologies have allowed robotics to substitute human functions in providing housekeeping, concierge, food and other service tasks (Cain et al. 2019; Ivanov and Webster2019; Yu 2020). This trend is generally not expected to be widely accepted since there is a desire to maintain high touchpoint amenities for tourism experiences (Zeng et al. 2020). Nonetheless, the digitalization of strategic aspects of tourism service does present opportunities to create value for destinations and their citizenry, and the pandemic is forcing people to accept touchless encounters in the interest of their health and safety.

Caribbean destinations possess some necessary foundational infrastructure in terms of internet technology and access, the major prerequisite for digital transformation in tourism. In Grenada, internet usage via computers, mobile phones and other digital technologies grew from 10,000 users in 2000 (GCDL 2020) to 56,000 users in 2017 for an estimated population of 111,219 (ISPs 2020). Progressive countries and leaders in digital technology, such as China, have demonstrated that integrating technology in a country's economic development agenda is one strategy toward diversification, but it requires policy intervention to provide direction. The rise of business ecosystems in China has been facilitating diversification using internet technology for finance, healthcare, culture and entertainment, business and enterprise services, among several other industries (Greeven and Wei 2017). China's lead in the development of multifaceted businesses (ecosystems) was greatly assisted by its government's deliberate policy interventions that sought to integrate internet technology into traditional industries thereby encouraging innovation, improving overall business efficiency, and ultimately facilitating rapid growth (Greeven and Wei 2017). A total of six policy interventions in 2015, which included the Internet Plus Policy, were instrumental in creating its dynamic business environment and internationalization (Greeven and Wei 2017).

The imperative to rebuild more sustainable ecologies and economies in the Caribbean post-pandemic in a manner that fosters resilience and re-imagines the tourism industry (Sheller 2020), may well require a new approach that is capable of resetting Caribbean tourism for the future. This may be implemented using some of the fundamental principles of collaborative economies (Botsman and Rogers 2010; Sundararajan 2016), diverse economies (Gibson and Graham 2008) as well as alternative development paradigms using technology as the impetus for sectoral diversification in these small economies.

Research Context

Grenada is one of the most southern tourism-dependent Caribbean destinations. Similar to other Caribbean countries, the post-colonial period was agrarian based but currently the economy is dominated by services (Thomas-Francois et al. 2017). In 2019, travel and tourism contributed

USD$ 59 billion to the Caribbean region (WTTC 2020b). The sector's contribution to the region's Gross Domestic Product (GDP) represented 13.9% of the total (WTTC 2020b), making the Caribbean the most tourism-dependent region in the world. In Grenada, travel and tourism contributes 10.3% to its economy's total GDP with 1 in 4 net new jobs created by the sector over the past five years (WTTC 2020a). The sector also provides 42.9% of the country's employment and 79.3% of total exports (WTTC 2020a).

The population of Grenada is relatively young with a quarter (24%) under the age of 25, while 90% of the population is below age 65 (National Plan Secretariat 2019). However, 37% of the population live below the poverty line with a high incidence of youth poverty among whom two-thirds are considered to be the working poor (National Plan Secretariat 2019). The country acknowledges the opportunities of rapid technological developments for national development, but also noted the weakness of a limited pool of highly technical skilled workers in the economy (National Plan Secretariat 2019). In addition, low value-added production exists on the island and as a result exports are low. Despite the country's usual resilience to disasters and international economic crises, an unprecedented shock such as COVID-19, which has been affecting travel and tourism for an extended period of time, has been catastrophic, requiring an infusion of new development strategies that are more robust and sustainable.

Research Methodology

This study used a combined research design which involved canvassing multiple secondary sources for relevant information that would inform the topic and pre-identified themes. These themes respond to the broad question of how can island destinations such as Grenada be more resilient post-pandemic and during repetitive external economic shocks. Information sources included a special issue of *Tourism Planning and Development* (2018, Issue 5) on Diverse Economies, geared toward exploring alternative economic models; articles from *Tourism Geographies* (May 2020) which addressed the way forward for the travel and tourism industry post-pandemic, industry articles specifically related to the Caribbean, and tourism related webinars, in particular "Reimagining Tourism in Post-pandemic Small Island Developing States (SIDS)" held on September 1, 2020 and co-hosted by the Alliance of Small

Island Developing States, the United Nations Foundation and the United Nations Development Programme, and "The Digital Revolution: Tools and Tactics to Understanding Today's Traveler" held on September 29, 2020 hosted by Skift. The content from these sources was coded within the themes; tourism product and service enhancements or substitutes; development of business ecosystems and tourism; new and existing markets; consumer service customization; and new modes for accessing destination marketing intelligence.

Using NVIVO Qualitative Research Software the data was stored and organized into the various themes. The findings also include observations of researchers based on current practices and trends in travel and tourism. This research strategy of employing content analysis allowed the study to be informed by multidisciplinary scholarship and real-world discourse through a review of relevant information, experiences and developments worldwide (Tight 2017). Content analysis is a method for summarizing any form of content by measuring repetition based on the focus of the study in order to draw conclusions from the content (Audience and Business Research 2014). The method permits classifying both written or oral materials into identified categories or themes of similar meanings (Moretti et al. 2011). The unobtrusive technique facilitated data collection without direct contact with the various sources during the peak of the COVID-19 crisis. The general strength of this research methodology is the ease of access to information that provides an understanding of social reality or phenomena through verbal or written communication materials, and its ability to allow coding of the data based on the relevant existing theory or previous research (Cho and Lee 2014). However, it is also noted for its labor intensiveness and is consequentially time consuming (Cho and Lee 2014). As posited by Audience and Business Research (2014), with this methodology it is impossible to be comprehensive and so the researchers' practical experience from different countries and cases together with observation also informed the research findings.

Research Findings and Discussion

In this section, we present the findings on the following five subtopics, the foundation of the thematic areas investigated in content analysis of the discourse in the literature and real-world context in re-imagining, diversifying and resetting tourism and travel post-pandemic. The findings are also presented with discussions and relevant supported cases or suggested strategies that are context specific to the tourism destination Grenada.

Tourism Product and Service Enhancements or Substitutes

For small developing destinations that are dependent on tourism, there is consensus that the travel and tourism industry needs to be rebuilt as carbon neutral and resilient. The strategies proposed for resilience emphasized diversification using existing assets in the destination, both human and physical, as a means of creating a more sustainable development model. These included embracing the Blue Economy (defined by the World Bank (2020) as the sustainable use of ocean resources for economic growth and improved livelihoods) with a focus on the surrounding ocean for renewable energy, developing Climate Smart Agriculture while also increasing agricultural production, and targeting high value/low volume tourism markets.

There is recognition of the need for the tourism product to be more diversified and the introduction of new tourism experiences such as leisure fishing, maritime archaeology, and virtual exploration. Greater involvement of communities is expected to be a source of accessing new ideas that are beneficial to resident populations. There is a push for involvement by local residents in product development using existing technology and social media to interact, educate and engage with potential future travelers. Rethinkers proposed advancing digital infrastructure for the creation of interactive platforms within the destination's communities to advance virtual tourism. Infrastructure includes community internet kiosks and universal access to broadband internet. The introduction of applications (apps) to guide visitors to tourism sites while in the destination or abroad would allow for co-creation between residents and visitors or potential visitors. The importance of the destination's communities and visitors or potential visitors' interaction and continuous connection via internet technology is described as a tool for creating empathy and emotional connections.

These new strategies, however, would require the destination to build new skills and capacities in the existing labor force. There must be targeted efforts to skill, upskill or reskill workers. It requires a concerted effort to train its youths as well as older workers in information and other innovative technologies and will require investments that must facilitate community entrepreneurships. Similarly, there is a need for training in climate smart agriculture techniques and energy production. One example is the Climate Smart Agriculture (CSA) pilot project funded by the Deutsche Gesellschaft fürInternationale Zusammenarbeit (GIZ) GmbH, Germany's leading international cooperation initiative for services. It has contributed significantly to awareness and provided some level of training to young and existing producers in Grenada. The initiative targeted the integration of agriculture development with climate responsiveness which seeks to protect natural resources that are also assets of the tourism sector.

Development of Business Ecosystems and Tourism

The need to create dynamic linkages through the development of new business ecosystems that are culturally appropriate emerged from the findings of the re-imagined travel and tourism sector. Starting with players in the destination's tourism value chain, but including non-tourism related micro, small and medium enterprises (MSMEs) (private sector), there must be effort to identify areas to collaborate in order to create diverse models of economic exchanges. For example, international airlines can be targeted to export not only tourists to the destination, but agro-processed and locally manufactured goods ordered by past visitors, diaspora communities and international consumers. The Belmont Estate's "My Grenada Box" epitomizes an example of targeting diaspora communities in the international market (Belmonte Estate 2020). The initiative started December 2020 during the pandemic, with a focus only on the US market but quickly expanded in scope to Europe and Canada. This initiative allows small agro-processing businesses in Grenada to package specialty foods and non-food items desired by the diaspora to be ordered by subscription from Grenada via the company's website (http://www.mygrenadabox.com). The boxes are then shipped to the United States, European and Canadian markets. The logistics for packing and distributing the boxes are assumed by the Belmont Estate, a long-standing agri-tourism business. In addition to the business that is generated for the participating agro-processors, a contribution from the sale of each box is donated to a charitable organization in Grenada.

The first step in establishing a connection with international markets for MSMEs will be to establish online businesses, a strategy used by China to develop very complex virtual business ecosystems(Greeven and Wei 2017). In the case of China, three independent platforms were built that facilitated online and offline shopping which diversified into different sectors (Greeven and Wei 2017). The most diversified and popular platform of its kind is Alibaba. The internet platforms should be able to support a wide range of businesses, including cultural entertainment, gaming, virtual tours and onsite destination shopping for goods and services facilitated online. This "lift and shift" type of innovation(Brooker and Joppe 2014) requires evidence-based policy development and implementation with a focus on establishing partnerships that are geared toward empowering local business people. Regional collaboration on the concept may also provide an impetus for faster international publicity.

New and Existing Market

There is a group of "new travelers" emerging from existing markets. Grenada's primary source markets may continue to be the United Kingdom, the United States and Canada. However, among the younger generations of travelers–mostly GenZ (generation z) and some Millennials–there tends to be greater concern for environmental protection, social justice and authenticity as well as sustainable touristic experiences that go beyond sand, sea and sun. It is therefore necessary that tourism product development include climate responsible practices and projects in all sectors that will present opportunities for this type of positioning. This includes the emphasis on agro-ecological and regenerative agricultural technologies that will help to improve food sovereignty and thus reduce the tourism food demand-related imports. This will also be beneficial to the local producers, create employment for youths and provide tourists with the desired local food experience (Thomas-Francois et al. 2017).

The cruise industry has been challenged as one source of severe environmental degradation. The large ships that can carry thousands of guests also often overwhelm small communities with the sheer number of passengers that disembark. Experts agree that a best practice will be to target smaller cruise ships that bring greater value and are designed to maximize cruise tourism benefits. These niche cruises use smaller vessels that offer differentiated experiences usually focused on educating visitors and cultural experiences. This therefore presents an opportunity for communities to be involved in the product development at the destination as well as opportunities for provisioning. There is the opportunity to link the destination value propositions, themes and product offerings to the experiences anticipated at the destination (for example, chocolate and spice tours) as part of the differentiated service offerings by the cruise line. Another benefit of targeting smaller cruise ships is related to reducing the number of visitors, and as a result this reduces the risk of carrying capacity concerns in fragile eco-systems in communities. Technology can be used to virtually display their wonders even when the tourists are onsite which will serve to protect communities' assets.

Globally, the tourism industry is preparing for an aggressive relaunch post-pandemic. There are also opportunities to connect with innovative technological initiatives that will spring forward. One such initiative is Livn API, an automated reservation system that allows resellers to connect to thousands of tours and activities suppliers. An open connectivity hub, it synergies industry players, harmonizes systems and infrastructures, merges and streamlines a traditionally fragmented tours and activities space and connects tour operators to travel resellers in real-time (Skift 2020, 2). This type of platform will allow tourists to spontaneously make travel arrangements since bookings and cancellations will be seen in real time. The technology will also allow for different channels and verticals (hotels, airlines, etc.) as well as access through different mediums, such as desktop, tablet, mobile phones and the travel agency environment (Skift 2020, 3). This infrastructure is well positioned to revolutionize travel, especially for younger travelers and provide MSMEs in the destination the opportunity to reach a global audience. It is important that Grenada's Chamber of Commerce and other small business development institutions point young emerging entrepreneurs to position themselves to engage with these new connectivity hubs. There are prospects for creating new business models that will foster greater independence in visitor sourcing as an alternative to overreliance on travel agents.

Consumer Service Customization

One of the most valuable assets of any destination is its people. Building capacity and awareness of the population to be custodians of tourism products and services, and mindfulness of the importance of providing highest quality services to all visitors in the destination is important. The perspective of accommodation facilities should be to extend highest service quality that serves to encourage repeat visits to their properties by tourists over extended periods. This was evident in an exemplary case study of a luxury resort in Grenada(Thomas-Francois et al. 2017). This study revealed that a service-oriented approach to local food service at the resort with a focus on its intangible resources, especially human resource capabilities, valued relationships, and value co-creation and production with customers and all members of the resort's service operations, allowed the resort to create and maintain competitive advantages. This type of operation led by local leadership, without resorting to expatriate employees, resulted in a consumer-centric high-end service offering. It was facilitated by interactions among members of the chain which produced continuous food service innovation that satisfied guests and was integral in their decision to repeat visits to the resort. This type of operation also presents an opportunity for youth's involvement in the agri-food system; however, a larger extent of youth involvement will require deliberate policy interventions to develop the necessary platforms that will permit youth participation to a greater scale (Thomas-Francois et al. 2017, 94).

There is the opportunity for resorts and hotels to offer customized service experiences to each guest using computer technology to store their profiles as was the case in the luxury resort. Computer technologies and information systems are useful in recording and responding to guests' profiles and preferences, including food, places to visit and preferred activities. Information technology can also allow accommodation properties to stay connected with guests after their visit and seize the opportunity to establish and create virtual communities for past visitors. Historically, crises such as COVID-19 have resulted in the introduction of technological innovations that are able to improve operational efficiency and effectiveness. Information systems geared toward improving service quality will assist accommodation properties to track and respond to guest needs. However, this must be done in concert with the built capacity and distinctive capabilities of employees to render consistent and high-quality consumer-centric services.

The Grenada Tourism Authority (GTA) in collaboration with the Central Statistical Office (CSO) has recently invested in Tourism Intelligence Solution (TIS) developed by the local company Sonover, led by young Grenadians. The web-based application was built for simplifying the collection, storage and analysis of data related to persons visiting Grenada, Carriacou & Petite Martinique. The data TIS is used for official tourism statistical, planning and marketing purposes by the GTA and CSO. TIS allows for efficiency in the collection of arrival data, capturing visitor profiles by source markets. Currently, Sonover provides the benefits of interactive information dashboards, automated assessments of tourism metrics and ease of tourism reporting which also informs marketing strategy (Sonover 2021). The next step would be the possibility for data mining for target marketing to existing and potential visitors and the use of artificial intelligence and recommender systems to predict needs based on behavior and purpose to create memorable customer-centric experiences for visitors.

New Modes for Accessing Destination Marketing Intelligence

Destination management organizations have adopted different innovative modes to promote tourism and attract potential tourists. Those practices provide useful implications for the marketing and sustainable development of Caribbean destinations. One of the practices is to apply gamification to tourism and hospitality marketing. Destination Ontario launched several gamified advertisements, such as *Where Am I*, *Ontario Colourful Spring Tour, and Fantastic Ontario Family Trip.* Instead of showing the landmarks and attractions of the province, the advertisements invited viewers to participate in games and then provided travel information. The gamified advertisements successfully attracted audiences, increased advertising engagement, and enhanced brand image. Another example is from Tourism New Zealand, which has launched an interactive online game *PLAY NZ* to show off New Zealand as the ultimate place to visit (Green 2020). The brand-building campaign targets Australian Millennials, and it was estimated that the travel bubble between Australia and New Zealand could contribute up to $3.7 billion to the New Zealand economy (Platt 2020). Grenada could also consider video games as a way to motivate potential tourists since experiencing virtual attractions significantly influences affection, impacts cognition, and triggers people's visit interest (Dubois and Gibbs 2018; Shen and Joppe 2021). Another good practice is to use social media (e.g., Blogs, Microblogs and content communities) as strategic platforms for connecting with travel writers, journalists, influencers and potential tourists. Destination Canada not only takes advantage of social media to reach international audiences, but also to monitor users' activities to discover their preferences and develop tourism products (Destination Canada 2016). Grenada should also utilize social media to share exciting stories, show off beautiful landscapes, and raise brand awareness. Google Analytics and Google Alerts could be used to monitor users' activities on the social media platforms of Caribbean destinations, such as Facebook, Twitter and Instagram. The data will give insights into how long users spend time there, what information is most attractive, and if activity converts into sales. Additionally, Grenada should consider using TikTok to promote destinations and attract young generations. As suggested by Van Dyke (2020), it is time for travel and tourism brands to get on TikTok and reach key audiences.

Conclusion

Caribbean destinations' heavy dependence on tourism over the decades has proven to be detrimental especially in times of unprecedented global health crises and economic collapse. It is now critical for island destinations such as Grenada to begin an aggressive diversification program that reduces the vulnerability of excessive dependence on the tourism sector. The integration of digital technology into the sector can be used as a conduit to branch off into other service sectors such as international trade of goods and services with the use of virtual platforms. This can further stimulate other sectors of the economy by creating demand for locally produced goods and services. One corresponding effect would be the demand for training and opportunities for youth employment in information and digital technology.

As postulated by Sheller (2020, 11), building backward and forward linkages between agro-ecological projects and local food markets together with a renewed vision for sustainable tourism that is non-extractive but supports local farmers, renewable energy micro-grids and more regenerative circular economies that reduce waste should constitute the rethinking of tourism's future in Grenada and other Caribbean destinations. This new mindset and way of delivering high-quality tourism experiences requires strategic planning and policy with forceful and guided direction. It goes without saying but government investments and leadership are critical to reposition, re-image and reset the industry to be resilient and sustainable in all senses of the word. As part of negotiating landing right and establish travel routes for international airlines to the destinations, it is customary for countries in the Caribbean to invest significant sums of money to guarantee seats to lure airlines to the island. Similarly, foreign direct investors in accommodation properties have enjoyed concessions and tax-free benefits not only during the years of construction but also for many years later. These approaches ought to be reviewed and negotiation strategies need to include priorities that are geared toward diversifying the economy as discussed in this chapter. A proactive and new approach that goes beyond CSR and that is also collaborative is necessary during negotiations with international tourism value chain players. Grenada is therefore in Phase 1 of the resilience adaptive cycle, a place reserved for planning out strategies that will engender innovation and creativity. The Caribbean will remain a desired and preferred place of visit for many source markets once it maintains its tranquility, beauty and safety. Valuing this fact, should be empowering during the process of a reset. In addition, an integrated technology approach to building resilience is a necessary and worthy investment into a future for economic prosperity. Technological transformation can help in truly allowing destinations to experience–All the world's a stage!

References

Amoamo, M., K. Ruckstuhl, and D. Ruwhiu. 2018. Balancing Indigenous Values Through Diverse Economies: A Case Study of Māori Ecotourism. *Tourism Planning & Development* 15 (5): 478–495.

Antoine , P., T. Taylor, and M. Church. 2014. *Grenada's Growth and Poverty Reduction Strategy (GPRS) 2014–2018*. Grenada: Government of Grenada

Ashley, C., and G. Haysom. 2006. From Philanthropy to a Different Way of Doing Business: Strategies and Challenges in Integrating Pro-poor Approaches into Tourism Business. *Development Southern Africa* 23 (2): 265–280.

Audience and Business Research. 2014. Know Your Audience: Chapter 16 Content Analysis. Retrieved from http://www.audiencedialogue.net/kya16a.html.

Benjamin, S., A. Dillette, and D.H. Alderman. 2020. "We Can't Return to Normal": Committing to Tourism Equity in the Post-pandemic Age. *Tourism Geographies*: 1–8.

Belmonte Estate. 2020. My Grenada Box. Retrieved from https://m.facebook.com/belmontestategd/videos/847736482725118/?refsrc=https%3A%2F%2Fm.facebook.com%2F110369408975324%2Fposts%2F3899675713377989%2F&_rdr.

Botsman, R., and R. Rogers. 2010. What's Mine Is Yours. *The Rise of Collaborative Consumption*.

Brooker, E., and M. Joppe. 2014. Developing a Tourism Innovation Typology: Leveraging Liminal Insights. *Journal of Travel Research* 53 (4): 500–508.

Cain, L.N., J.H. Thomas, and M. Alonso, Jr. 2019. From Sci-fi to Sci-fact: The State of Robotics and AI in the Hospitality Industry. *Journal of Hospitality and Tourism Technology*.

Cave, J., and D. Dredge. 2018. *Reworking Tourism: Diverse Economies in a Changing World*. Taylor & Francis.

Cho, J.Y., and E.-H. Lee. 2014. Reducing Confusion about Grounded Theory and Qualitative Content Analysis: Similarities and Differences. *The Qualitative Report* 19 (32): 1.

Clayton, A., and N. Karagiannis. 2008. Sustainable Tourism in the Caribbean: Alternative Policy Considerations. *Tourism and Hospitality Planning & Development* 5 (3): 185–201.

CTO. (2020). Grenada Reports Stellar Tourism Performance in 2019 [Press release]. Retrieved from https://www.onecaribbean.org/grenada-reports-stellar-tourism-performance-in-2019/.

Destination Canada. 2016. Building Capacity for International Media Relations: A Toolkit for Canadian Tourism Businesses. Retrieved from https://www.destinationcanada.com/sites/default/files/2016-11/Resources_MediaRelationsToolkit_Nov2016_EN.pdf.

Dubois, L.E., and C. Gibbs. 2018. Video Game-Induced Tourism: A New Frontier for Destination Marketers. *Tourism Review* 73 (2): 186–198.

GCDL. 2020. Our World in Data, Sustainable Development Goals Tracker : Grenada Internet Usage Retrieved from https://ourworldindata.org/grapher/number-of-internet-users-by-country?tab=chart&country=~GRD.

Gibson-Graham, J. 1996. *The End of Capitalism (as we knew it): A Feminist Critique of Political Economy*. Oxford: Blackwell.

Gibson-Graham, J.K. 2006. *A Postcapitalist Politics*. University of Minnesota Press.

Gibson, K., and J. Graham. 2008. Diverse Economies: Performative Practices for 'other Worlds' and Diverse Economies in Geography Online Bibliography. I Had Corrected These Gibson-Graham References for you, but See You Didn't Accept Them. Why?

Greeven, M.J., and W. Wei. 2017. *Business Ecosystems in China: Alibaba and Competing Baidu, Tencent*. Xiaomi and LeEco: Routledge.

Green, R. 2020. Tourism New Zealand Invites Gamers to Play NZ In World-Style Gaming Experience Via Tbwa\Sydney. Retrieved from https://campaignbrief.com/tourism-new-zealand-invites-gamers-to-play-nz-in-world-style-gaming-experience-via-tbwasydney/.

Gyimóthy, S., and J.W. Meged. 2018. The Camøno: A Communitarian Walking Trail in the Sharing Economy. *Tourism Planning & Development* 15 (5): 496–515.

Hughes, E., and R. Scheyvens. 2018. Development Alternatives in the Pacific: How Tourism Corporates Can Work More Effectively with Local Communities. *Tourism Planning & Development* 15 (5): 516–534.

ISPs. 2020. Grenada Internet Service Providers (ISPs) List. Retrieved from https://www.newsmedialists.com/isp/grenada I corrected this one as well.

Ivanov, S., and C. Webster. 2019. Conceptual Framework of the Use of Robots, Artificial Intelligence and Service Automation in Travel, Tourism, and Hospitality Companies. In *Robots, Artificial Intelligence, and Service Automation in Travel, Tourism and Hospitality*. Emerald Publishing Limited.

Lew, A.A., J.M. Cheer, M. Haywood, P. Brouder, and N.B. Salazar. 2020. Visions of Travel and Tourism after the Global COVID-19 Transformation of 2020. *Tourism Geographies*: 1–12.

McLeod, M., E.H. Lewis, and A. Spencer. 2017. Re-inventing, Revolutionizing and Transforming Caribbean Tourism: Multi-country Regional Institutions and a Research Agenda. *Journal of Destination Marketing & Management* 6 (1): 1–4. https://doi.org/10.1016/j.jdmm.2016.08.009.

Moretti, F., L. van Vliet, J. Bensing, G. Deledda, M. Mazzi, I. Rimondini, and Fletcher. . 2011. A Standardized Approach to Qualitative Content Analysis of Focus Group Discussions from Different Countries. *Patient Education and Counseling* 82 (3): 420–428.

National Plan Secretariat. 2019. *National Sustainable Development Plan (NSDP) 2020–2035*. Grenada: Published by the National Plan Secretariat.

Platt, C. 2020. Play NZ Campaign Launches with Loserfruit as Tourism New Zealand Targets Australian Gamers. Retrieved from https://www.traveller.com.au/play-nz-campaign-launches-with-loserfruit-as-tourism-new-zealand-targets-australian-gamers-h1pnnw.

Rhiney, K. 2016. From Plantations to Services: A Historical and Theoretical Assessment of the Transition from Agrarian to Service-based Industries in the Caribbean. In *Globalization, Agriculture and Food in the Caribbean*, 23–50. London: Palgrave Macmillan.

Rifkin, J. 2014. *The Zero Marginal Cost Society: The Internet of Things, the Collaborative Commons, and the Eclipse of Capitalism*. St: Martin's Press.

Skift. 2020. *4 Future-Proof Reasons for Travel Companies to Automate Tours and Activities in 2021 and Beyond*. Published Livn + Skift- Nov 24, 2020 Retrevied from: https://skift.com/2020/11/24/livn-future-proof-reasons-travel-companies-automate-tours-activities-2021/?utm_campaign=Daily.

Sonover. 2021. Grenada Tourism Authority: Unleashing Destination Data. https://sonover.com/work/grenada-tourism-authority.

Sheller, M. 2020. Reconstructing Tourism in the Caribbean: Connecting Pandemic Recovery, Climate Resilience and Sustainable Tourism Through Mobility Justice. *Journal of Sustainable Tourism*: 1–14.

Shen, Y., and M. Joppe. 2021. Gamification: Practices and Challenges from Stakeholders' Perspectives. In *Gamification for Tourism*, eds. F. Xu and D. Buhalis. Channel View Publications.

Sundararajan, A. 2016. *The Sharing Economy: The End of Employment and the Rise of Crowd-Based Capitalism*. Cambridge, MA and London: MIT Press.

Statista. 2020a. Grenada: Number of Foreign Visitor Arrivals 2010–2019, Published by Ana M. López, October 21, 2020. Retrived from https://www.statista.com/statistics/816408/grenada-number-of-tourist-arrivals/#:~:text=Grenada%2C%20a%20Caribbean%20country%20located,inbound%20arrivals%20in%20both%20years.

Statista. 2020b. Grenada: Internal Travel and Tourism Consumption 2010–2019, Published by Ana M. López, October 21, 2020. Retrived from https://www.statista.com/statistics/814785/grenada-tourism-revenue/#:~:text=Grenada%3A%20internal%20travel%20and%20tourism%20consumption%202010%2D2019&text=In%20total%2C%20Grenada's%20consumption%20of,approximately%20490%20million%20U.S.%20dollars.

Thomas-Francois, K., M. Von Massow, and M. Joppe. 2017. Service-oriented, Sustainable, Local Food Value Chain—A Case Study. *Annals of Tourism Research* 65: 83–96. https://doi.org/10.1016/j.annals.2017.05.008.

Tight, M. 2017. *Understanding Case Study Research: Small-scale Research with Meaning*. Sage.

Van Dyke, A. 2020. It's Time for Travel and Tourism Brands to Get on TikTok. Retrieved from https://beautifuldestinations.com/blogs/news/travel-brands-on-tiktok.

World Bank. 2020. What Is the Blue Economy? Retrieved from https://www.worldbank.org/en/news/infographic/2017/06/06/blue-economy.

WTTC. 2020a. Grenada 2020 Annual Research: Key Highlights. Retrieved from https://wttc.org/Research/Economic-Impact.

WTTC. 2020b. Travel and Tourism Regional Performance in 2019. Retrieved from https://wttc.org/Research/Economic-Impact.

Yu, C.-E. 2020. Humanlike Robots as Employees in the Hotel Industry: Thematic Content Analysis of Online Reviews. *Journal of Hospitality Marketing & Management* 29 (1): 22–38.

Zeng, Z., P.-J. Chen, & A.A. Lew. 2020. From High-touch to High-tech: COVID-19 Drives Robotics Adoption. *Tourism Geographies*: 1–11.

PART IV

Economic Resilience

CHAPTER 10

Increasing the Resilience of Micro, Small and Medium Tourism Enterprises to Tropical Cyclones in Small Island Developing States

Thalia Balkaran and David Smith

Introduction

Natural hazards such as earthquakes, tropical storms and associated flooding, droughts and volcanoes have led to social, economic and environmental destruction, losses to human life and livelihoods (UNISDR 2013). Disasters that occur as a result of these hazards disrupt the normal functioning of society and their impacts may have long-standing effects within a country. The Caribbean which is a hazard-prone region experienced 165 disaster events between 1990 to 2008 with damage and losses for this period being estimated at USD 136 billion (ACS 2017). Tropical Cyclones represent the most significant natural threat to coastal communities (Mousavi et al. 2011). They bring high-intensity rainfall,

T. Balkaran (✉) · D. Smith
Institute for Sustainable Development, The University of the West Indies, Kingston, Jamaica

D. Smith
e-mail: david.smith@uwi.edu

A. Lewis-Cameron et al. (eds.), *Managing Crises in Tourism*,
https://doi.org/10.1007/978-3-030-80238-7_10

heavy winds and storm surge. These storms may disrupt transportation networks, housing, critical facilities and vital sectors of the economy. These impacts can be especially destructive in Small Island Developing States (SIDS) that depend on coastal areas and resources. These large-scale disturbances have a profound impact on the economy, they are a major source of macroeconomic vulnerability with the most intense having the capacity to wipe out the GDP of an entire year. The region has seen catastrophic damage over the years, more recently the 2017 hurricane season left a path of destruction across many islands within the region. Hurricane Irma followed by Hurricane Maria left the islands of the north-east Caribbean in ruins, Barbuda and Dominica were hardest hit, both islands and British Virgin Islands according to the United Nations will require 5 US Billion to rebuild; Hurricane Maria erased decades of development for Dominica which was already one of the Eastern Caribbean's poorest nations with a poverty rate estimated at 28.9% (21,000 persons) (Moloney 2017).

Hazards tend to have a greater impact in SIDS due to a host of socio-economic, environmental and political factors that shape their vulnerability. These islands tend to be dependent on climate-sensitive sectors like tourism that have the potential to be wiped out with a single hazard event. Since the 1980s, tourism has distinctly improved in SIDS, acting as an engine of growth which provides an important source of foreign exchange and jobs directly and indirectly through support sectors such as agriculture and transport (Craigwell 2007). The tourism industry is an important economic driver to the Caribbean region, in 2016 the sector contributed to 15.2% of GDP (56.4 million USD) and 13.8% of employment (2.4 million jobs) (WTTC 2018). The islands of the region however are exposed annually to tropical cyclones that threaten tourism and tourism livelihoods. This is of particular concern since climate change is expected to increase the severity of these events.

Most of the businesses within the tourism industry are small or medium enterprises that provide a range of tourism services and products (Buhalis and Cooper 1998; Richards and Hall 2006; Becken and Hay 2012). Because of the importance of Micro, Small and Medium Tourism Enterprises (MSMTE's) to the economy of SIDS, it is important to focus disaster research on these entities and suggest measures to increase their resilience. Resilience is described as the ability of a community or society exposed to hazards to absorb, adapt and recover from an event in a timely manner through risk management (UNISDR 2017). The Hyogo

Framework for Action highlighted that the beginning point for decreasing disaster risk and building disaster resilience was through an understanding of "hazards and the physical, social, economic and environmental vulnerabilities to disasters that most societies face... followed by action taken on the basis of that knowledge" (UN 2005, 1). By understanding underlying vulnerabilities, this chapter is able to suggest disaster risk reduction (DRR) measures to increase the resilience of tourism livelihoods.

Literature Review

Disasters represent a growing risk for businesses within the private sector; catastrophes disrupt the normal business process directly by the impact on facilities and resources and indirectly by causing damage to infrastructure and public utilities such as transport and energy networks that businesses depend on (UNISDR 2013). These types of businesses are especially vulnerable to any form of serious disruption to their financing, human resources or supply chain (Bittle 2018). Long-term competitiveness and sustainability can be affected when business is interrupted, business may be lost to another competitor, skilled employees may leave to find other jobs and reputation permanently lost (UNISDR 2013). Businesses once lost may never return and some businesses never recover following a disaster (UNISDR 2013). Small and medium enterprises are more likely to be more vulnerable since they have less knowledge, resources and risk-reducing schemes to prepare and respond to events (Cioccio and Michael 2007; Wang and Ritchie 2012; UNISDR 2013). This is in contrast to large businesses and global corporations who tend to be more resilient due to their greater investment in DRR, good insurance coverage, diversity and operations in many countries which buffers the impact of events (UNISDR 2013).

There are studies focusing on the vulnerability of SIDS to disasters (Pelling and Uitto 2001; Méheux et al. 2007; Le Masson and Kelman 2011; Jackson et al. 2017), but only a few have focused on the vulnerability of tourism to disasters within SIDS (Méheux and Parker 2006; Mahon et al. 2013; Becken et al. 2014) despite its importance to the economy of most SIDS. Further, none of these studies emphasized the vulnerability of livelihoods to hurricanes. Only a handful of research papers (Cioccio and Michael 2007; Baker and Coulter 2007; Calgaro and Lloyd 2008) focused on tourism businesses and livelihoods within the context of vulnerability; however these studies focused on fires, terrorism or tsunamis respectively.

This study attempts to fill that gap by studying the vulnerability of tourism livelihoods to tropical cyclones in two SIDS. Many studies that focus on the vulnerability of livelihoods tend to do so in reference to climate change. These studies tend to highlight fishing and agricultural livelihoods. Little research has been carried out on how livelihoods interplay with a person's vulnerability within the context of natural hazards (Gaillard et al. 2009). There has been research carried out by Gaillard et al. (2009) which examines livelihoods and vulnerability within the context of natural hazards, however this is done within the fishing sector. This study differs from previous studies because it focuses on the vulnerability of tourism livelihoods to tropical cyclones (hurricanes and tropical storms) by examining the socio-economic and environmental factors contributing to vulnerability. This information is then used to suggest measures to build the resilience of tourism MSME's.

Methodology

Information presented here is part of a larger mixed methods study conducted on the vulnerability of MSMTE's to tropical cyclones. For this chapter, emphasis is placed on the qualitative aspect of the study. Two islands were selected to give a more holistic understanding of MSMTE vulnerability as it relates to tropical cyclones across the region. Tourism is a critical economic activity in both islands contributing to GDP, employment and foreign exchange. It is the importance of tourism for both countries combined with the risk both islands face to hurricane activity that led to their selection. There are also differences in the tourism sector in each island that would bring contrasting themes to the study. Coomansingh (2004) describes tourism in the twin-island state as being in an embryonic state compared to Jamaica where there is a well-developed tourism sector.

The methodology used in this research allowed for an examination of vulnerability at the community level, this approach allowed for the understanding of complex local realities at each site. It is important to take into consideration local perceptions since it leads to a deeper understanding of community vulnerability and the formulation and success of disaster mitigation measures, strategies and policies (Nirupama 2012). Four coastal sites (Figs. 10.1 and 10.2) were chosen across both islands based on

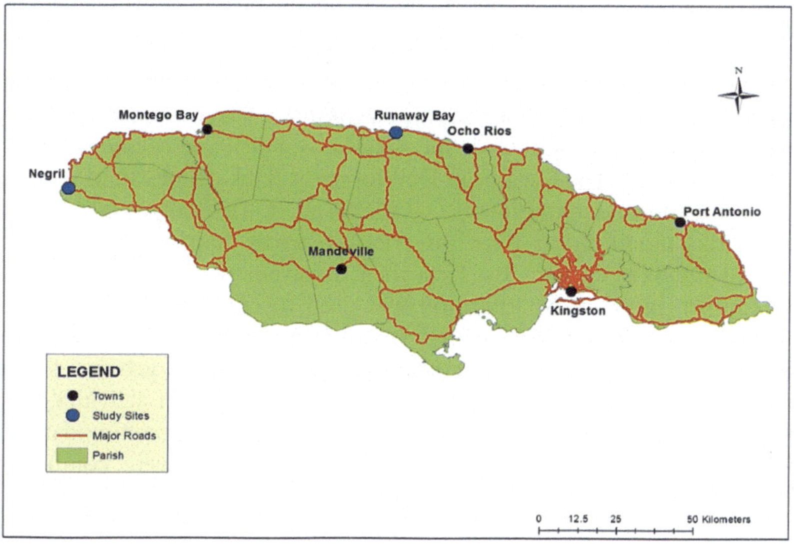

Fig. 10.1 Location of Negril and Runaway Bay in relation to other towns within Jamaica (*Source* Authors)

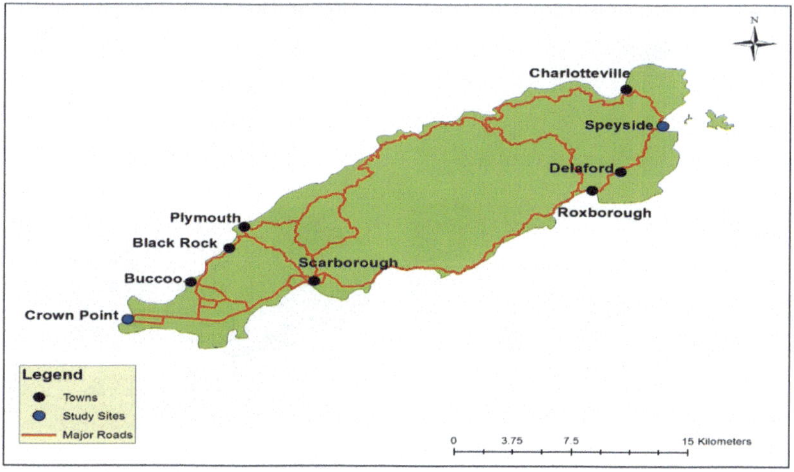

Fig. 10.2 Location of Crown Point and Speyside in relation to other towns within Tobago (*Source* Authors)

their importance within the tourism industry and their size which allowed interviews to be carried out in a timely manner.

Qualitative interviews were conducted to collect in-depth information on MSME's at each site. For the purpose of this research employment was used as the categorization to define the Micro, Small and Medium Enterprises (MSME's). The Caribbean Development Bank defines an MSME as having 50 and under employees (CDB 2016). A wide range of MSME's fitting this classification were interviewed across tourism (Fig. 10.3). Participants were chosen based on their involvement and knowledge of the tourism industry. Stratification was used to create varying strata or groups within the population under study, this allowed the research to highlight differences noticed between varying livelihoods groups. There were three noticeable groups that stood out as the main activities that persons from the tourism industry were involved in: Accommodation, Craft and Watersports/Tours. Watersports and Tours were combined since many of the persons involved in watersports also carried out land or

Fig. 10.3 Pictures showing an example of MSMTE's interviewed for this research Upper left: A glass-bottom boat in Crown Point, Upper Right: Hotel offering accommodation in Speyside, Bottom Left: Craft hut in Speyside, Bottom Right: Craft enterprise in Runaway Bay. (*Source* Balkaran, T.)

water tours and vice versa. The author decided that the research would be reported taking into consideration the viewpoints of persons belonging to these sectors of the tourism industry.

Although the focus of the study was persons who worked directly within tourism, representatives from other organizations that were not a part of the industry were interviewed due to their knowledge or potential to inform on issues related to tourism. This included participants from the private sector, government agencies, NGO's and other stakeholders with an interest in tourism such as representatives from relevant ministries, environmental agencies and groups, livelihood associations, disaster agencies and tourism industry representatives. Prior research and reconnaissance trips were carried out at each site by the author. This helped to identify places of interest relevant to the study and to familiarize potential interview participants with the author and the purpose of the research. At the onset of the qualitative fieldwork, purposive sampling, a form of nonprobability sampling was used to target persons for interviews. Patton (2002, 242) in explaining this type of sampling refers to the selection of what he calls "information-rich cases" which are individuals that are strategically and purposefully selected to learn the most on issues of importance. Due to the stratification created, stratified purposeful sampling was used to sample persons across the three main groups at each site. In this type of sampling, the sampling frame is divided into fairly homogeneous subgroups and then samples are purposely selected for each stratum (Onwuegbuzie and Collins 2007). Interviewees were also asked to refer to other persons like themselves who could be interviewed (snowballing). This assisted in gaining access to participants in the population under study.

Interviews were carried out in the study area and within each group until a repetition of responses was observed and it was determined by the interviewer that no new insights were being gained. The author determined the sample size and point of saturation based on the necessary information required to complete the study. Sixty-three interviews were conducted for Negril, 20 for Runaway Bay, 31 for Crown Point and 20 for Speyside. Table 10.1 shows a breakdown of interviews conducted at each site.

The first site where interviews were carried out was Negril. This was carried out as part of project by The Partnership for Canadian Caribbean Climate Change Adaptation (PARCA) which focused on coastal livelihoods in study sites in the Caribbean and Canada (PARCA

Table 10.1 A breakdown of the numbers of interviews conducted at each site

Group	*Negril*	*Runaway Bay*	*Crown Point*	*Speyside*
Accommodation	15	7	5	3
Craft	20	6	4	3
Watersports/ Tours	13	4	11	9
Key informants	15	3	11	5
Total	63	20	31	20

n.d). The project conducted a Community-Based Vulnerability Assessment (CBVA), the author was a part of the research team that conducted this fieldwork from the 19th November to 19th December 2012. Permission was gained to use part of the information collected towards this research. The CBVA used a guide to conduct semi-structured interviews that focused on topics covering tourism, disasters, livelihoods and socio-economic and environmental issues affecting the community. Although questions were provided for each category, interviews were conducted in a conversational style with the author at times diverging away from the guide and editing questions to suit each individual interview and the investigation as time went by. This research methodology was then replicated in the other study sites. Fieldwork for Tobago and Runaway Bay was undertaken by the author in 2015. Qualitative content analysis, a common approach used in analyzing qualitative documents which searches for themes in the material through the development of categories (Bryman 2004), was used as the method of interpreting the text from the interviews.

Findings and Discussion

There were common themes and factors that emerged from the qualitative analysis. Factors common to both countries affecting the vulnerability of MSMTE's to tropical cyclones included: "closeness to the sea, the lack of insurance and access to loans, increased competition within the tourism sector, the seasonality of tourism and the level of support through governments and social networks." This section presents succinct risk-reduction measures that can be implemented from the "top down" to the "grass roots" level to increase livelihood security and resilience to tropical cyclone events. These suggestions were made based on the main themes,

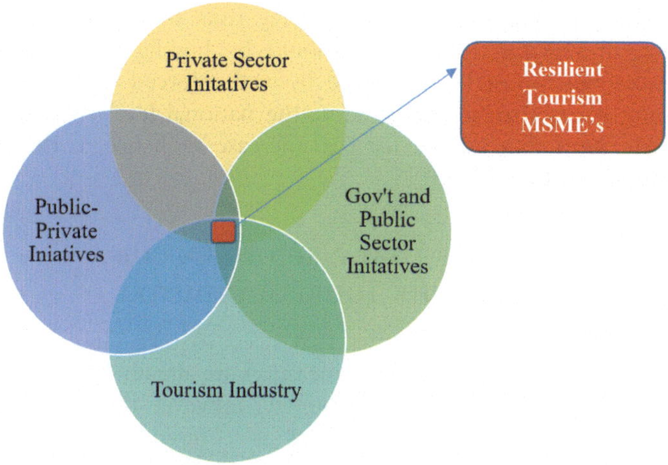

Fig. 10.4 Resilient tourism businesses produced by the combination of measures from 4 main actors (*Source* Authors)

conclusions drawn by the author and expressed needs and gaps identified by stakeholders. These suggestions are intended to be a template of measures that can be used to increase resilience within MSMTE's across small islands. Resilient tourism MSME's can be produced with the combination of initiatives suggested across the four main actors (Fig. 10.4).

Private Sector Initiatives

The business case for investment in DRR is particularly vital in SIDS where the private sector investment in the economy is high; In countries where enforcement and legislation is weak, private sector initiatives can lead DRR efforts (Mahon et al. 2013). According to the Sendai Framework, the lack of incentives and regulation of DRR measures by the private sector is an underlying risk factor, these underlying risk factors should be addressed through disaster risk-informed private and public investments which contribute to sustainable development and are more cost effective than relying on post-disaster response and recovery (UNISDR 2015). The recommendations suggested for businesses place emphasis on Prevention and Preparedness. Businesses that have the best practices of managing their property risks have a better chance at long-term survival, however it should be highlighted that private sector investment must be reinforced by well-enforced regulations such as building codes that are enforced at the national level (Mahon et al. 2013). Measures that can be used to increase resilience in the private sector are shown in Table 10.2.

Changes that Need to Occur Within the Tourism Industry

This section examines the changes that need to take place in tourism based on the analysis. Some of these measures are already being addressed in some form in each respective country, however the recommendations here represent what the author saw as required needs in the study sites. A recommendation made does not mean that the country is not taking steps towards addressing these issues. The extent to which some of these measures have been implemented is not within the scope of this study. When questioned about the current state of tourism, there was an overwhelming tone of negative responses as a result of what respondents stated was declining tourism and profits over the years. This was more commonly associated to Micro-Businesses (Craft and Watersports), although there were some businesses in Accommodation that expressed difficulties as well. In Jamaica large-all-inclusive hotels were blamed for taking away the business from MSME's while in Tobago there was a sharp decrease in international tourism due to what respondents described as a lack of focus on the tourism industry by the central government which is located in Trinidad. Despite being located in a tourism-intensive region, Trinidad and Tobago and particularly Trinidad is dominated by the energy sector which has been the mainstay of the economy and has encouraged growth since the oil boom in the seventies (Lewis and Jordan 2008). This success has diverted attention away from the tourism sector and government policies have not favoured the development of the industry (Lewis and Jordan 2008).

Tourism as an industry is subjected to fluctuations. Interviewees described what they referred to as "ups and downs" over the years which some persons attributed to the unpredictable nature of operating a tourism business. Financial changes or difficulties represent a challenge as explained by Adger (2006, 270), "vulnerability of livelihoods to shocks occurs when people have insufficient real income and wealth, and when

Table 10.2 Suggested measures to promote resilience in the private sector

Private sector measures
1. **Crisis Management Plans**
Across the sites only within the Accommodation group, was their mention of having a plan in place to deal with a hazard event, even within this group there were businesses who reported no such plan. Each MSME should have a crisis management plan that takes into consideration employees and where appropriate tourists, these plans should be updated regularly (Pennington-Gray et al. 2011)
2. **Business Continuity Management (BCM)**
It is critical for tourism MSMEs to have in place a Business Continuity Plan to ensure long-term survival and organizational resiliency following any form of disruptive event (Bittle 2018). In BCM and the creation of a business continuity plan, the following should be done:
a. a business continuity risk assessment to identify and assess the risks and vulnerabilities (NDPTC 2018)
b. Design a risk management programme through which risks will be dealt with. According to the NDPTC (2018), either through
• Avoidance—avoiding placement of business in high-risk areas e.g. Places prone to flooding,
• Loss control (loss prevention, loss reduction) e.g. back up data in a cloud
• Risk Financing—figuring out ways to pay for losses through

(continued)

Table 10.2 (continued)

Private sector measures
- Acceptance/Retention (using businesses' own funds) - Transfer (using external funds). This can be done through • **Insurance transfer**—it is important to understand what insurance covers, coverage may include that for infrastructure, equipment and property insurance (NDPTC 2018). It is also important to purchase the most appropriate policy based on finances and risk e.g. flooding insurance in a flood-prone region • **Noninsurance transfer**—e.g. Identify possible sources of disaster assistance programmes e.g. government assistance programs c. Training with employees to increase awareness of disasters/increase resilience e.g. "Hurricane Preparedness Week" (NDPTC 2018) d. Consider supply chain management - consider which materials, labour or supplies may be disrupted after a disaster, what difficulties a disaster will create in obtaining them and how to ensure the supply of these (NDPTC 2018). This will ensure a resilient supply chain (NDPTC 2018). This varies according to the business, hotels may need food and watersport and boat MSME's may need gas to start operating. There should be a plan for post-disaster supply (NDPTC 2018) e. Build Partnerships - Relationships should be built with customers and partners (other businesses, NGO's & governmental agencies). This builds social capital. In Puerto Rico after Hurricane Maria in 2017, a small watersports/restaurant business "Taino Divers/Shipwreck Bar and Grill" was able to set up a Go-Fund me account to help with associated costs after the disaster, the business was able to use their social capital and relationships with past customers to raise 41,150 of their 50,000 dollar goal (USD) in 12 months (NDPTC 2018; GoFundMe 2018) Very few MSME's interviewed had these plans in place. According to Bittle (2018) in the Caribbean, there is a general lack of awareness and training in BCM in MSME's. BCM and resilience planning requires funding for training and implementation, many MSME's however do not have the resources or expertise. Government agencies or non-profit agencies need to develop the necessary expertise to provide support to these businesses. The National Disaster Management offices can expand their role to include awareness and training efforts to inform businesses on natural hazards and BCM (Bittle 2018)

Private sector measures
3. **Insurance**
This risk transfer method reduces overall loss in the event of a disaster (Mahon et al. 2013). In many developing countries including SIDS insurance is not always readily available where insurance markets are limited or premiums too expensive for small businesses (Mahon et al. 2013). Many of the MSMEs in Tobago and Jamaica expressed difficulty as it related to insurance; the use of Index insurance can aid MSMEs. Index insurance pays a fixed amount based on experienced weather conditions regardless if losses were experienced (Hall et al. 2012). It is linked to an index e.g. rainfall, if this index is exceeded the insurance company makes a payment (Hall et al. 2012). The advantage of this approach is that it can target different scales from National Governments to individual businesses which make it advantageous for MSMEs that dominate tourism (Hall et al. 2012)
One such scheme already being utilized in the Caribbean region is the Livelihood Protection Policy (LPP) designed to inject quick liquidity to provide financial stability for low-income vulnerable persons. It is designed to stabilize the livelihoods of vulnerable persons such as tourism workers, farmers and day labourers (CCRIF 2018b). The LPP was developed by The Climate Risk Adaptation and Insurance in the Caribbean project implemented by the Munich Climate Insurance Initiative (MCII) in partnership with the Caribbean Catastrophe Risk Insurance Facility Segregated Portfolio Company (CCRIF SPC). CCRIF SPC offers tropical cyclones, excess rainfall and earthquake policies to the governments of Caribbean and Central American countries (CCRIF 2018a). With the LPP, payments are released to policy buyers within 7 days once an event occurs beyond the pre-defined trigger level. There is no need for policy makers to make a claim since payments are triggered when thresholds for wind and rain are exceeded; the amount paid is determined by the severity of the event so a more extreme event will trigger a larger payment (CCRIF 2018b). The maximum coverage available is 4000 USD, but more than one policy can be purchased. The advantage of the LPP is that there are no limits placed on annual income making it possible for low-income persons to be eligible (CCRIF 2018b). Under the first phase, the LPP was made available in 3 pilot countries, Jamaica, St. Lucia and Grenada. Phase II expanded access to the LPP in the 3 pilot countries as well as in Trinidad and Tobago and Belize (CCRIF SPC 2020)
4. **Increasing Self-Reliance**
Across the sites in Tobago and Jamaica and more so among the micro-businesses there seemed to be an over-reliance on the government for solving problems, even as it related to Disaster Risk Management (DRM). Bittle (2018) found that because many MSMEs do not have BCM plans and resiliency procedures they tend to rely on typical forms of disaster relief such as emergency government funding. Businesses must develop good business practices such as "pay the business first" so that cash reserves can be accumulated to cover their immediate needs after a disaster (Bittle 2018)

(continued)

Table 10.2 (continued)

Private sector measures
5. **Associations**
It is advantageous for MSME's to be part of a small business association such as a craft, watersport or hotel association. From the interviews, it was seen that through associations there was a collective representative voice for members. These MSMEs are part of a larger network of small businesses, because of this they should partner with like-minded business owners, cooperatives and industry associations to strengthen their collective efforts, this will allow businesses to work together for their common good (Bittle 2018). In order to reduce the dependence on governments that cannot always provide aid for everyone, these associations should establish an Emergency Fund where members pay a fee on a monthly basis. At the Negril Craft Market, the craft association put in place this fund where money can be used to cover any damages that may occur as a result of a disaster. After a period of time, the association can also decide to use these funds to upgrade buildings, invest back into businesses or use for marketing
6. **Construction, Upgrades and Retrofitting**
Construction standards should be adhered to, this is particularly important for the Accommodation Sector. Retrofitting allows for improvements to be made. This may include roof replacement using stronger material, the addition of hurricane straps, shutters and hurricane two ply glass (Mahon et al. 2013). Retrofitting will ensure that buildings have a better chance against hurricane force winds. Across the sites there needs to be upgrades and retrofitting done to ensure building structures are able to withstand storms and strong winds. Following an event, if damage is sustained, reconstruction must take into consideration the appropriate building codes

there is a breakdown in other previously held endowments." Some of the recommendations made in Table 10.3 are therefore related to aiding in the success of the business required for increasing profits. Profits could be used to afford insurance, retrofit or provide upgrades needed to increase tourist traffic and experience.

Table 10.3 Measures needed in the tourism industry to increase the resilience of MSME's

Measures within the tourism industry
1. Registration of Businesses
In both islands there were businesses (particularly the micro-businesses, some crafters and small hotels) that were not registered. Informality has restrictions for health and safety, employee work conditions and the selection of operating locations (MOLSMED 2014). It is therefore important for the governments to make registration attractive to MSMEs, particularly for micro-businesses (MICAF 2017)
2. Training, development and funding of MSMTE's
From the interviews across both islands, there was an identified need for better training in the tourism industry and customer service. This should be facilitated through the relevant associations and agencies responsible for tourism. Completing basic training in some of these principles should be necessary for granting a license to operate. The benefits of tourism must be spread equally to ensure sustainability, local communities must benefit socially and economically, one way of this is through the development and support of MSMEs (Ministry of Sport and Tourism 2001). Increased support through funding to support the development of MSMEs through aspects like loan financing and business support services should occur (Ministry of Sport and Tourism 2001).
3. Water Provision
Water scarcity was a relevant theme affecting more so Accommodation businesses in the larger sites of Negril and Crown Point in both islands. Considering this it is essential to address water woes; a combination of measures must be put in place:
i. Future planning and approvals particularly for all-inclusive hotels that host thousands of additional tourists must be taken into consideration. Planning should involve the relevant water authorities and future expansion and construction should take into consideration current water infrastructure
ii. Authorities need to put in place additional measures to deal with the increased demand around high season which coincides with the dry season and drought. While expensive, desalination may be one of the options that islands take into consideration to ensure water needs are met. Currently 10–15% of water supplies come from Desalination in Trinidad and Tobago (Tapper et al. 2011)

(continued)

Table 10.3 (continued)

Measures within the tourism industry
4. Environmental Degradation
The region's dependence on tourism means that coastal and marine resources that attract tourists here must remain in pristine condition to ensure that it is here for future generations and the success of the industry. While in both countries there was acknowledgement of efforts of prosecution, marine patrols and education, there still remains room for improvement. In both countries, there are relative pieces of legislation, policies, plans and frameworks as it relates to environmental protection of coastline, marine and protected areas, what is needed however is greater governmental, organizational and financial commitment to the protection of the environment. A key component of this will be increased enforcement as it relates to laws and tourism planning
5. Enforcement and updating of Building Codes
Where necessary these should be revised to address poor standards and improve engineering designs and construction methods (Lewsey et al. 2004; Mahon et al. 2013). The Caribbean region is known for tourism development in coastal regions, research has shown however that strict observance of building codes reduces the devastation associated with extreme events (Lewsey et al. 2004). Revision of building codes should take place to take into consideration the following: -Climate change and its possible effects e.g. sea-level rise -the placement of infrastructure and their elevation which is critical in hazardous areas and when taking into consideration expected changes in sea level rise -improvement in construction techniques with the proper roofing material being used in addition to hurricane straps -specific measures to ensure that when new buildings are constructed they can withstand flooding and wind speeds (Lewsey et al. 2004)
6. Funding Proposals
While governments may want to dedicate greater financial resources to environmental protection, the reality is that these funds are not always available. Governments may consider using tourism dollars to reinvest back into environmental protection. A fee should be collected from each tourist that enters the country to contribute to a conservation/environmental fund. Jamaica already uses tourism dollars to invest back into tourism through the Tourism Enhancement Fund which collects money from incoming tourists; the fund aims to provide sustainable tourism development, promote growth and better management of environmental resources and enhance the experience of tourists (Ministry of Tourism 2016)

Public/Private Sector Partnerships

Often DRR is spearheaded by public sector agencies, usually the Disaster Risk Management Office. Public–private sector partnerships involving government agencies, international donors and private sector agencies such as hotel associations should take place to ensure that DRR measures are incorporated by different sectors and groups in society. There were examples of partnerships in both islands, there should be continued commitments to existing partnerships and innovative partnerships that seek to enhance livelihood security. Businesses and/or Associations representing tourism workers should have regular contact with local disaster agencies. This research found that tourism workers were better prepared and educated on disaster events when a member of their association had regular meetings with the disaster agency. This can be through training sessions chaired by the disaster agency or at regular meetings where representatives have discussions with stakeholders on a regular basis. In Tobago, representatives from the hotel associations mentioned meetings on a regular basis that included tourism representatives, the police and the Tobago Emergency Management Agency (TEMA). In Jamaica representatives from associations also mentioned meetings and workshops chaired by the Office of Disaster Preparedness and Emergency Management (OPDEM). Partnerships involving the public sector should aim to assist businesses to assess the costs and benefits of Disaster Risk Reduction (DRR) (Mahon et al. 2013). Through workshops, opportunities to increase the technical knowledge of tourism operators should take place (Mahon et al. 2013).

Research (including that performed by academia) and outreach into the business case for resilience should be supported, research should focus on financial and economic data that would possibly support and encourage greater investment in private sector DRR (Mahon et al. 2013). Regional and sector-specific research should also be encouraged (Simpson and Gladin 2008). Participatory approaches to Disaster Risk Management have been recognized in many global initiatives including The Yokohama Strategy and Plan of Action for a Safer World (1994) which encourages community involvement and the use of traditional knowledge and methods supplemented by technological and scientific knowledge, a theme supported by The Sendai Framework for Disaster Risk Reduction (Carby 2018). Governments should adopt an approach supportive of local initiatives, these government-civil society partnerships can decrease the costs of risk reduction, cultivate planning and implementation of DRR and ensure local acceptance (UNISDR 2009).

Government and Public-Sector Initiatives

Governments in both islands must take charge in spearheading measures discussed in Table 10.3. Governments have a leading, regulatory and coordination role when it comes to DRM, governments should engage with the necessary stakeholders including poor persons, those with disabilities, women and practitioners within the field to design and implement suitable standards, policies and plans (UNISDR 2015). In addition to the measures discussed so far, other underlying problems outside of tourism must also be addressed. The Sendai framework suggests addressing underlying risk drivers such as poverty and inequality, policies that are not risk informed, weak institutional arrangements, climate change, poor land management, lack of incentives and regulation for private sector participation in DRR measures and unsustainable uses of natural resources (UNISDR 2015). Additionally "improving preparedness and national coordination for disaster response, rehabilitation and reconstruction, and post-disaster recovery and reconstruction to 'Build Back Better,' supported by strengthened modalities of international cooperation" is critical (UNISDR 2015, 10). The governments in both islands and other SIDS should focus on options that increase resilience across a range of hazards (Mahon et al. 2013). In order to increase resilience to storm events the private sector must invest in the protection of their assets; this together with governments providing the national policy framework in which private development occurs, will seek to reduce risk within tourism (Mahon et al. 2013).

Conclusion

This chapter has suggested measures to increase the resilience of MSMTE's to tropical cyclone events. Measures are not only within the realm of Disaster Management but improvements must be made within the tourism industry as a whole and at the level of individual businesses. The tourism industry operates at varying scales and involves different types of associations, organizations and businesses, the high level of horizontal and vertical integration as well as the importance of private and public entities therefore requires a multi-stakeholder approach to resilience (Mahon et al. 2013). In order to ensure the best operational practices required for DRR, there needs to be cross-sectoral coordination between sectors like tourism and DRM (Mahon et al. 2013). The government, the public as well as the private sector are all necessary stakeholders within the tourism and DRR process. With all the measures discussed so far, it must be underlined by including communities and local populations in developing risk management plans, there is enhanced risk reduction when multiple knowledge systems including local, indigenous, scientific and policies are incorporated (Simpson and Gladin 2008). The approach to disaster risk must be a preventative people-centred one which includes scientific and research institutions, academia and the public and private sector, businesses themselves should integrate disaster risk into their management practices (UNISDR 2015). Considerable effort will be required by governments, policymakers, the tourism industry and individual business owners to ensure sustainable and successful risk management to increase their resilience to disaster events.

References

ACS (Association of Caribbean States). 2017. *Natural Hazards in the Caribbean: Too Costly to Ignore*. Accessed March 23, 2020. http://www.acs-aec.org/index.php?q=disaster-risk-reduction/natural-hazards-in-the-caribbean-too-costly-to-ignore.

Adger, W. Neil. 2006. Vulnerability. *Global Environmental Change* 16 (3): 268–281. https://doi.org/10.1016/j.gloenvcha.2006.02.006.

Baker, Kathleen, and Alex Coulter. 2007. Terrorism and Tourism: The Vulnerability of Beach Vendors' Livelihoods in Bali. *Journal of Sustainable Tourism* 15 (3): 249–266. https://doi.org/10.2167/jost643.0.

Becken, Susanne, and John Hay. 2012. *Climate Change and Tourism: From Policy to Practice*. New York: Routledge.

Becken, Susanne, Roché Mahon, Hamish G. Rennie, and Aishath Shakeela. 2014. The Tourism Disaster Vulnerability Framework: An Application to Tourism in Small Island Destinations. *Natural hazards* 71 (1): 955–972. https://doi.org/10.1007/s11069-013-0946-x.

Bittle, Micheal 2018. The Development of Sector Specific Training Material for Mainstreaming Gender-Sensitive Comprehensive Disaster Management into Business Practice for Small and Medium Size Enterprises. Report prepared for the Enhancing Knowledge and Application of Comprehensive Disaster Management (EKACDM) Initative.

Bryman, Alan. 2004. *Social Research Methods*, 2nd ed. New York: Oxford University Press.

Buhalis, Dimitrios, and Chris Cooper. 1998. Small and Medium Sized Tourism Enterprises at the Destination. In *Embracing and Managing Change in Tourism: International Case Studies*, ed. Bill Faulkner, Eric Laws, and Gianna Moscardo, 324–345. London: Routledge.

Calgaro, Emma, and Kate Lloyd. 2008. Sun, Sea, Sand and Tsunami: Examining Disaster Vulnerability in the Tourism Community of Khao Lak, Thailand.

Singapore Journal of Tropical Geography 29 (3): 288–306. https://doi.org/10.1111/j.1467-9493.2008.00335.x.

Carby, Barbara. 2018. Integrating Disaster Risk Reduction in National Development Planning : Experience and Challenges of Jamaica. *Environmental Hazards* 17 (3): 219–233. https://doi.org//10.1080/17477891.2017.1415864.

CCRIF (Caribbean Catastrophe Risk Insurance Facility). 2018a. About Us. Accessed 28th September, 2018. https://www.ccrif.org/content/about-us.

———. 2018b.*Livelihood Protection Policy (Lpp)*. Accessed 28th September, 2018 https://www.ccrif.org/projects/crai/livelihood-protection-policy-lpp.

———. 2020. Course Manual Fundamentals of Disaster Risk Financing for Advancing Sustainable Development of Small Island Developing States (Sids). The University of the West Indies, Open Campus.

CDB (Caribbean Development Bank). 2016. Micro-Small-Medium Enterprise Development in the Caribbean Towards a New Frontier. Accessed March 1, 2017. https://www.caribank.org/publications-and-resources/resource-library/thematic-papers/micro-small-and-medium-enterprise-development-caribbean-towards-new-frontier.

Cioccio, Leah, and Ewen J. Michael. 2007. Hazard or Disaster: Tourism Management for the Inevitable in Northeast Victoria. *Tourism Management* 28 (1): 1–11. https://doi.org/10.1016/j.tourman.2005.07.015.

Coomansingh, Johnny. 2004. The Nasty Side of Tourism Development: An Example from Trinidad and Tobago. *E-Review Tourism Res* 2: 15–21. https://agrilifecdn.tamu.edu/ertr/files/2012/09/145_The_nasty_side_of_tourism_development.2.15.2004.14-20.pdf.

Craigwell, Roland. 2007. Tourism Competitiveness in Small Island Developing States. Vol. 2007/19: Research paper 2007/19, UNU-WIDER, United Nations University (UNU).

Gaillard, Jean-Christophe, Emmanuel A. Maceda, Elodie Stasiak, Iwan Le Berre, and Maria Victoria O. Espaldon. 2009. Sustainable Livelihoods and People's Vulnerability in the Face of Coastal Hazards. *Journal of Coastal Conservation* 13 (2–3): 119. https://doi.org/10.1007/s11852-009-0054-y.

Go Fund Me. 2018. *Shipwreck Bar & Grill & Taino Diver*. Accessed October 6, 2018. https://www.gofundme.com/shipwreckbar-grill.

Hall, C. Michael., Stefan Gössling, and Daniel Scott. 2012. *Tourism and Climate Change: Impacts, Adaptation and Mitigation*. London: Routledge.

Jackson, Guy, Karen McNamara, and Bradd Witt. 2017. A Framework for Disaster Vulnerability in a Small Island in the Southwest Pacific: A Case Study of Emae Island, Vanuatu. *International Journal of Disaster Risk Science*: 1–16. https://link.springer.com/article/10.1007/s13753-017-0145-6.

Lewis, Acolla, and Leslie-Ann. Jordan. 2008. Tourism in Trinidad and Tobago: Carving a Niche in a Petroleum-Based Economy. *International Journal of Tourism Research* 10 (3): 247–257. https://doi.org/10.1002/jtr.654.

Lewsey, Clement, Gonzalo Cid, and Edward Kruse. 2004. Assessing Climate Change Impacts on Coastal Infrastructure in the Eastern Caribbean. *Marine Policy* 28 (5): 393–409. https://doi.org/10.1016/j.marpol.2003.10.016.

Mahon, Roché, Susanne Becken, and Hamish G. Rennie. 2013. Evaulating the Business Case for Investment in the Resilience of the Tourism Sector of Small Island Developing States : A Background Paper Contributing to the Global Assessment Report on Disaster Risk Reduction (Gar) 2013. New Zealand : Lincoln University. https://www.preventionweb.net/english/hyogo/gar/2013/en/bgdocs/Mahon%20et.al.%202012.pdf.

Masson, Le., and Virginie, and Ilan Kelman. . 2011. Disaster Risk Reduction on Non-Sovereign Islands: La Réunion and Mayotte, France. *Natural Hazards* 56 (1): 251–273. https://doi.org/10.1007/s11069-010-9566-x.

Méheux, Kirstie, Dale Dominey-Howes, and Kate Lloyd. 2007. Natural Hazard Impacts in Small Island Developing States: A Review of Current Knowledge and Future Research Needs. *Natural Hazards* 40 (2): 429–446. https://doi.org/10.1007/s11069-006-9001-5.

Méheux, Kirstie, and El. Parker. 2006. Tourist Sector Perceptions of Natural Hazards in Vanuatu and the Implications for a Small Island Developing State. *Tourism Management* 27 (1): 69–85. https://doi.org/10.1016/j.tourman.2004.07.009.

MICAF (Ministry of Industry, Commerce, Agriculture and Fisheries). 2017. *Jamaica Micro, Small and Medium Enterprises and Entrepreneurship Policy.* http://www.miic.gov.jm/sites/default/files/pdfs/MSME%20and%20Entrepreneurship%20Policy%20Updated%202017_0.pdf.

Moloney, Anastasia. 2017. *Donors Pledge $2 Billion to Help Hurricane-Hit Caribbean Recover*. Accessed May 5, 2018. https://www.reuters.com/article/us-caribbean-hurricane-aid/donors-pledge-2-billion-to-help-hurricane-hit-caribbean-recover-idUSKBN1DM2CN.

MOLSMED (Ministry of Labour and Small and Micro Enterprise Development). 2014. Micro and Small Enterprise (Mse) Development Policy for Trinidad and Tobago 2014–2016 Driving Entrepreneurship. Accessed October 5, 2018. https://www.ilo.org/dyn/youthpol/en/equest.fileutils.dochandle?p_uploaded_file_id=547.

Mousavi, Mir Emad, Jennifer L. Irish, Ashley E. Frey, Francisco Olivera, and Billy L. Edge. 2011. Global Warming and Hurricanes: The Potential Impact of Hurricane Intensification and Sea Level Rise on Coastal Flooding. *Climatic Change* 104 (3–4): 575–597. https://doi.org/10.1007/s10584-009-9790-0.

NDPTC (National Disaster Preparedness Training Center). 2018. Small Business Disaster Resilience. Power Point Presentation by National Disaster Preparedness Training Center at PRIMO Conference 2018.

Nirupama, Niru. 2012. Risk and Vulnerability Assessment: A Comprehensive Approach. *International Journal of Disaster Resilience in the Built Environment* 3 (2): 103–114. https://doi.org/10.1108/17595901211245189.

Onwuegbuzie, Anthony J., and Kathleen M.T. Collins. 2007. A Typology of Mixed Methods Sampling Designs in Social Science Research. *The Qualitative Report* 12 (2): 281–316. https://files.eric.ed.gov/fulltext/EJ800183.pdf.

PARCA. n.d. About Parca. The Partnership for Canadian Caribbean Climate Change Adaptation. Last Modified 2016. Accessed September 1, 2016. http://parca.uwaterloo.ca/about/.

Patton, Michael Q. 2002. *Qualitative Research and Evaulation Methods*, 3rd ed. California: Sage Publications.

Pelling, Mark, and Juha I. Uitto. 2001. Small Island Developing States: Natural Disaster Vulnerability and Global Change. *Environmental Hazards* 3: 49–62. https://doi.org/10.3763/ehaz.2001.0306.

Pennington-Gray, Lori, Brijesh Thapa, Kyriaki Kaplanidou, Ignatius Cahyanto, and Elaine McLaughlin. 2011. Crisis Planning and Preparedness in the United States Tourism Industry. *Cornell Hospitality Quarterly* 52 (3): 312–320. https://doi.org/10.1177/1938965511410866.

Richards, Greg , and Derick Hall. 2006. Conclusions. In *Tourism and Sustainable Community Development*, ed. Greg Richards and Derick Hall, 297–306. London: Routledge.

Simpson, M.C., and E. Gladin. 2008. *Good Practices: Natural Hazard Risk Management in the Caribbean Tourism Sector*. Barbados: Caribbean Tourism Organization. https://www.onecaribbean.org/content/files/NaturalHazardRiskManagementFinaltext.pdf.

Tapper, Richard, Michalis Hadjikakou, Rachel Noble, and Joseph Jenkinson. 2011. The Impact of the Tourism Industry on Freshwater Resources in Countries in the Caribbean, Mediterranean, North Africa and Other Regions. London, Uk: Travel Foundation.

The Ministry of Tourism and Sports. 2001. *Master Plan for Sustainable Tourism Development*. London: Commonwealth Secretariat. Accessed May 24, 2014. http://www.tpdco.org/wp-content/uploads/2016/07/Master-Plan-for-Sustainable-Tourism-Development.pdf.

The Ministry of Tourism. 2016. *Tourism Enhancement Fund*. Accessed July 16, 2017. http://www.mot.gov.jm/agency/tourism-enhancement-fund.

UN (United Nations). 2005. Hyogo Framework for Action 2005–2015:Building the Resilience of Nations and Communities to Disasters. Extract from the final report of the World Conference on Disaster Reduction (A/CONF.

206/6).Accessed January 12, 2019. https://www.unisdr.org/files/1037_hyogoframeworkforactionenglish.pdf.

UNISDR (United Nations International Strategy for Disaster Reduction). 2009. Summary and Recommendations Global Assessment Report on Disaster Risk Reduction Risk and Poverty in a Changing Climate Invest Today for a Safer Tomorrow. Geneva: United Nations. https://www.preventionweb.net/files/9414_GARsummary.pdf.

———. 2013. Global Assessment Report on Disaster Risk Reduction 2013, the Pocket Gar 2013 from Shared Risk to Shared Value: The Business Case for Disaster Risk Reduction. France: United Nations. https://www.preventionweb.net/english/hyogo/gar/2013/en/home/GAR_pocket/GAR_pocket_3.html.

———. 2015. *Sendai Framework for Disaster Risk Reduction 2015–2030*. Geneva: United Nations. http://www.preventionweb.net/files/43291_sendaiframeworkfordrren.pdf.

———. 2017. *Terminology*. Accessed March 30, 2018. https://www.unisdr.org/we/inform/terminology.

Wang, Jie, and Brent W. Ritchie. 2012. Understanding Accommodation Managers' Crisis Planning Intention: An Application of the Theory of Planned Behaviour. *Tourism Management* 33 (5): 1057–1067. https://doi.org/10.1016/j.tourman.2011.12.006.

WTTC (World Travel and Tourism Council). 2018. *Caribbean Resilience and Recovery: Minimizing the Impact of the 2017 Hurricane Season on the Caribbean's Tourism Sector*. https://wttc.org/Research/Insights/moduleId/1000/itemId/12/controller/DownloadRequest/action/QuickDownload.

CHAPTER 11

Building Resilience by Strengthening the Link Between Tourism and Agriculture: An Assessment of the Purchasing Patterns of Selected Hotels and Guesthouses in Jamaica

Tolulope Bewaji, Eritha Huntley Lewis, Clive Scott, and Twila-Mae Logan

Introduction

Tourism is considered to be imperative to the economic growth of developing nations (Torres and Henshall Momsen 2004). This is because it is a strong source of foreign direct investment as well as foreign exchange earnings. Over the last decade, there has been a consistent increase in the

T. Bewaji (✉) · E. H. Lewis · C. Scott
The University of the West Indies, Kingston, Jamaica
e-mail: tolulope.bewaji02@uwimona.edu.jm

E. H. Lewis
e-mail: eritha.huntley@uwimona.edu.jm

C. Scott
e-mail: clive.scott@uwimona.edu.jm

A. Lewis-Cameron et al. (eds.), *Managing Crises in Tourism*,
https://doi.org/10.1007/978-3-030-80238-7_11

key tourism indicators for Jamaica, i.e., tourism arrivals and tourist expenditure. In 2010, for example, despite the effects of the Global Recession, Jamaica recorded total arrivals of over 2.9 m visitors and received tourism expenditure of US$2billion. By 2019, the island received 4.2 m tourists, an increase of 45% in total tourist arrivals over 2010 and earned over US$3.6 billion, an increase of 80% over 2010 tourism expenditure. In 2019, there was a 15.9% decrease in cruise passengers partially offset by an 8.4% increase in stopover visitors, resulting in an overall decrease of 2% over 2018 arrivals (Jamaica Tourist Board 2019).

The tourism sector's contribution to GDP has increased from 6.9% in 2010 to 9.8% in 2019 at market prices (Statistical Institute of Jamaica n.d.) but negatively affects the balance of payment since so much food is imported to sustain the tourism sector. The finding of the Tourism Demand Study (Ministry of Tourism & Entertainment 2015) for Jamaica states that leakage due to agricultural imports was J$1.6billion (US$1 = J$117.31). At the same time however, significant profits from the tourism sector are repatriated back to the headquarters of the hotel company. For resilience to exist, a successful backward linkage is important in the tourism industry.

The nature to the tourism industry, therefore, presents an excellent opportunity for the growth and development of other economic sectors through the creation forward and backward linkages. The more the linkages experienced with the tourism industry, the better the economic picture of a country (Ntibanyurwa 2006). For Jamaica, tourism has a direct and multiplier effect on the economy. For example, the money spent in the hotel is used by the hotel to get resources from the farmers. The farmers then use the funds to get seedlings, fertilizers, etc. These vendors will further use the money to procure good and services not related to tourism (Wilson 2020).

Tourism and Agriculture in Jamaica

Sir Arthur Lewis, Nobel Prize winner with expertise in developing countries, outlined an economic growth plan for Caribbean, Asian and African

T.-M. Logan
Chaplin School of Hospitality & Tourism Management, Florida International University, Miami, FL, USA

countries decades ago. He articulated that '(t)he farmers' position is much more hopeful if development begins outside agriculture ... This, in turn, generates an increase in demand for agricultural products, and so development spreads from sector to sector' (Lewis 1983). Lewis believed that industrialization goes along with agricultural development and that the agricultural sector should never be ignored. He advocated for agriculture but unfortunately, his prescription for economic growth was not utilized in its entirety in the Caribbean (Lewis 1983).

Agriculture is important to the success of the Jamaican economy. The agricultural sector recorded a 9.4% annual compounded growth rate between 2010 and 2019, and the sector's contribution to GDP has increased from 5.2% in 2010 to 7.0% in 2019 at current prices (Statistical Institute of Jamaica n.d.). Strengthening the link between tourism and agriculture through the creation of opportunities for backward and forward linkages is critical to the development of a resilient economy. According to Lewis (1983; cited by (Figueroa 2004)), the subsistence sector needs to be transformed for the development of other industries. Fortunately, efforts are now being made to increase the amount of local agriculture products that funnels into the tourism sector. One such effort in Jamaica is the Tourism Agri-linkages Project online trading platform, Agri-Linkages Exchange (ALEX), which is funded and supported by Tourism Enhancement Fund, Ministry of Tourism and Rural Agricultural Development Authority RADA. The goal of this interface is to strengthen the linkages between agriculture and tourism sector, ultimately resulting in increased domestic production and reduction in the importation of agricultural goods. This represents the type of cross-industry collaboration that is critical for overall economic development (RADAAdmin n.d.).

During the completion of this book chapter, the entire world was engulfed in the COVID-19 pandemic. As a result, the tourism sector suffered significantly. In Jamaica, where borders were closed from April to June 15th, 2020, stopover arrivals declined by 66.3% between January 2020 and October 2020 (Jamaica Tourist Board 2020). The economic fallout will necessitate that the industry takes appropriate actions to emerge stronger from this crisis. The tourism industry in Jamaica is stated to be a resilient one with a history of overcoming disruptions successfully (Wilson 2020). Resilience is defined as 'the capacity of a system to absorb disturbance and reorganize while undergoing change so as to still retain essentially the same function, structure, identity, and feedbacks' (Walker et al. 2004). Just as the US auto and financial industries needed to survive the 2008 crisis, Jamaica's economy needs tourism to survive because of its direct and multiplier effect. This industry needs to survive disturbances and reorganizes to survive successfully. Further, to recover from COVID-19, the most resilient economies will be driving the recovery, and travel and tourism will be a multiplier—and an employment engine across all sectors (Wilson 2020).

This chapter, therefore, examines the nature of the tourism-agriculture relationship in the context of Jamaica to ascertain the strength of that relationship as the need for resilience and recovery becomes more critical. It is proposed that the main impetus for the resilience of the tourism industry is the success of the linkage between these two sectors. This is explored by asking the following questions:

Research Questions.

1. What is the nature of the relationship between the tourism and agriculture in Jamaica?
2. What are the major sources of agricultural produce for accommodation establishments in Jamaica?
3. Is the size of the accommodation establishment a determinant of their purchasing behavior?
4. What are the main challenges (if any) facing accommodation establishments in procuring agricultural produce from local farmers?

Literature Review

Theoretical Framework

The multiplier theory was first developed in the nineteenth century and formalized by Keynes and other economists in the 1930s. The multiplier theory explains how the injection of money into the economy can result in increased economic activity beyond the initial injection of expenditure by visitors. Numerous studies have used this theory to demonstrate the importance of inflows from tourist expenditure generating sustainable economic growth especially for developing countries (Mathouraparsad and Maurin 2017; Khan et al. 1995; Kweka et al. 2003). The essential elements of this theory are shown in Fig. 11.1.

The multiplier effect increases economic activity in three ways. Direct effects occur as visitors make payments for goods and services—hotel

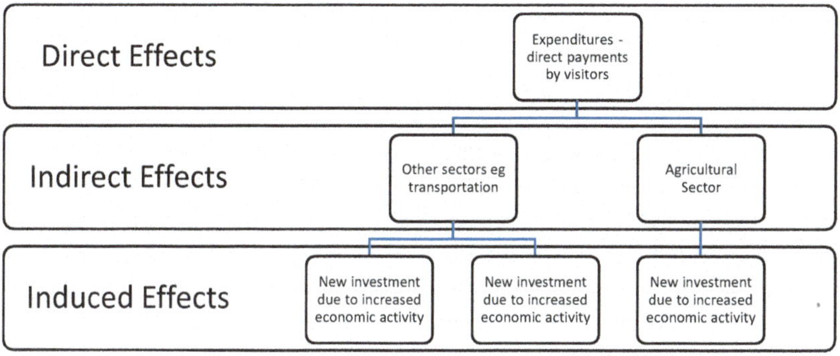

Fig. 11.1 Multiplier Effects

rooms, food and beverage, souvenirs, etc. Indirect effects are realized when payments are made to related sectors; for example, agricultural goods are purchased from local farmers. Finally, induced effects are as a result of new investments in unrelated sectors, for example new housing development for hotel workers or new business openings that provide increased demand for goods and services arising from the income earned from the indirect effects.

The size of the multiplier effect is impacted by leakages. Leakages occur due to two factors (i) savings as individuals generally save a portion of their income and (ii) importation. Leakages due to importation of goods can significantly weaken the multiplier effects reducing the full potential of the tourism sector to provide the requisite boost to economic growth. For example, if the tourism sector were to import all the goods and services used in this sector, then economy would enjoy only the direct effects and there would be no indirect or induced effects (Khan et al. 1995).

The multiplier effect is dependent on the strength of the linkages between tourism and the other sector especially the agricultural sector. Linkage is an inter-industry relationship that involves moving backward or forward on the value chain of activities. Forward linkage deals with output utilization and backward linkage deals with the relative importance of a sector as a consumer of input from other sectors (Anderson 2018). Backward linkage in terms of agriculture is important as it facilitates one of the biggest expenditures within the tourism industry. Successful backward linkage leads to an increase in the multiplier effect and lowers the foreign exchange leakage that occurs due to the tourism industry acquiring foreign food items (Torres 2002). Successful forward linkage is seen to be correlated with strong economic growth (Hsieh and Kung 2013). The tourism industry has a strong potential for backward and forward linkage (Kweka et al. 2003).

A theory that is used in the linkage analysis is that of value chain approach. The value chain approach is an operational way of looking at the production process with a goal of having successful integration of activities along the value chain to increase productivity and development (Mitchell and Ashley 2009). It allows one to determine the beneficiaries and losers in the value chain from production to final market. It [also] allows one to evaluate the interactions along the value chain of activities (Anderson 2018). It is an important tool for studying intersectoral linkages in important economic sectors. The value chain approach has been used to analyze the linkages in the tourism industry in developing countries (Lacher and Nepal 2010; Mitchell and Ashley 2009). The tourism sector in Jamaica is very important because of its direct and the far-reaching linkages that it has with other sectors of the economy (Anderson 2018). Thus, the linkage between agriculture and tourism allows for tourism to be more economically inclusive (Sims 2008). For example, research shows that more and more tourists are interested in the food of the locale that they are visiting (Sims 2008). Experiencing local food makes the trip more authentic and it boosts the sustainability of the industry.

The multiplier effect results in tourism creating jobs not only in the tourism industry but in other industries (Trebicka 2016). The combination of successful linkages as well as the multiplier effect can be sufficient to protect a developing nation like Jamaica from the most adverse effects of the COVID-19 pandemic. Linkages promote the sustainability of the tourism industry as local relationships are built that can reduce the negative impacts of external crises.

Linking Tourism and Agriculture

The tourism sector has the potential, through the multiplier effect, to increase economic growth by forming stronger linkages with the agricultural sector which has shown significant growth in the last 10 years (Statistical Institute of Jamaica n.d.). There is also high level of receptivity to the idea of import substitution which would increase production and could reduce the dependence on imports significantly (Ministry of Tourism & Entertainment 2015). This receptivity for import substitution presents a tremendous opportunity for trade that could be as high as J$56.7 billion annually for processed foods alone, J$5.3 billion for fruits and J$1.6 billion (US$1 = J$117.31) for vegetables (Ministry of Tourism & Entertainment (MTE) 2015). Of note, Jamaica has been one of the few Caribbean countries to cut their imports leakage from around 69% in the late 1960s (Belisle 1983) to between 8.5 and 25.5% in recent years (MTE 2015, 11; cited by (Deere and Royce 2019). This may potentially result in a positive outlook for the tourism-agriculture relationship in Jamaica with both sectors benefiting from this collaboration.

This tourism-agriculture relationship, though beneficial, is sometimes fraught with challenges as problems of local access sometimes outweigh the benefits. Hotels still need to somehow gain provisions for their kitchen and guests. (Torres and Henshall Momsen 2004) found that in Cancun the chefs stated significant issues that lowered their interest in local produce. They found, at times, the local production was non-existent, of low quality, irregular supplies, and that prices are high. Worse, tourists do not always desire the local produce (Torres and Henshall Momsen 2004). There were also insufficient efforts by tourist planners to ensure that agriculture development strategies are created (Torres 2003).

In Nepal, the research shows that there are benefits from linking tourism with areas such as poultry and dairy (Shah and Gupta 2000). These authors also found that when formal tourism enterprises are owned by locals, they are more likely to engage in backward linkages by using local suppliers. Similarly, in Crete, most of the tourism businesses, especially the smaller enterprises, were more likely to purchase from local suppliers (Andriotis 2002).

Further research into the linkage between agriculture and hotels in various regions in the world provides a more nuance view of how middlemen can address some of the supply and demand challenges (Kang and Rajagopal 2014; Torres and Henshall Momsen 2004). In Indonesia, middlemen were used to facilitate getting the produce from farmers and fishermen to the hotels. Though middlemen were successful in resolving some supply chain issues in fishery, they had limited success with farmers (Torres and Henshall Momsen 2004).

Evidence from Uganda shows that simple food items such as fruits and vegetables are bought by the hotels from intermediary suppliers or farmers on the local market (Adiyia and Vanneste 2018). However,

Ugandan hotels saw inconsistencies in the quality and quantity available locally in comparison with the standards required by foreign tourists (Adiyia and Vanneste 2018). Items such as protein, pasta, rice and spices tended to be sourced internationally, from supermarkets or middlemen when local supplies are not of sufficiently high standards (Adiyia and Vanneste 2018). An interesting purchasing behavior occurs in Uganda where hotels let their staff members to act as intermediaries and choose the suppliers. If these suppliers are deemed trustworthy—quality of the produce is good—then the hotel will maintain a relationship with the supplier (Adiyia and Vanneste 2018).

Smaller and non-starred hotels should be included in research on linkages as these vary based on the size of organizations (Mshenga and Richardson 2013). It was found that four- and five-star hotels had a strong linkage with local supply networks while the smaller hotels rely solely on local food supplies (Telfer and Wall 1996). In a further, it was found that in Cancun hotels, 84 and 85% of vegetables and fruits, respectively, were gained from specialized wholesalers (Torres 2003), while, as noted earlier, meat is mostly imported (Torres 2003). In Tanzania, having cooperatives developed by the farmers was considered advantageous for improved linkage between the farmers and the hotels (Nguni 2016). It allows them to produce and market together and thus, gain economies of scale (Nguni 2016). This author suggests that by working together to sell, *and* to produce, helps to deal with issues of quality, quantity and consistency.

Roy et al. (2017) found that trust was important for successful linkages between restaurants and farmers. If there is trust, restaurants are more likely to continue buying from these farmers, but this can result in new farmers having difficulties accessing these restaurants (Roy et al. 2017). They also found in addition to trust, the desire for value and locality of the business affected purchasing behavior. Even though this study used restaurants, it is expected that these results would also be applicable to hoteliers.

Deere and Royce (2019) show that Cuba's *paladares* [private restaurants] strengthened tourism-agricultural linkages in two ways: through farm-to-table arrangements within the expanding agroecological farming sector and by the steady demand for high quality produce largely supplied by Cuba's small private farmers. The researchers point out that scale was critical in this tourism-agriculture relationship, thus hotels, even when well-intentioned, end up relying on middlemen for fresh food who procure it at local or regional wholesale markets rather than engaging in farm-to-table arrangements (Deere and Royce 2019).

This chapter, therefore, explores the interactions along the value chain of activities. The lessons learnt can be used to increase the economic inclusiveness of the tourism industry and increase the resilience and sustainability of the tourism industry.

Methodology

This research was conducted using a quantitative approach. This approach was selected for the following reasons: (i) the data required for this exploratory study could be readily obtained from closed-end questionnaires in the form of numerical scores, e.g., ranking using a Likert scale, and (ii) data collected in numerical form is easily presented in tables and charts, using percentages and means, appropriate for inductive, exploratory and descriptive analyses commonly used in leisure and tourism research (Veal 2018).

The study population comprised 343 licensed accommodation establishment/properties that offered food and beverage service according to the Jamaica Tourist Board list of active establishments in 2015 (Jamaica Tourist Board 2015). The properties were divided accordingly: 181 hotels and 161 guesthouses. This list was obtained from the Tourism Product Development Company Ltd (TPDCo), the entity tasked with the responsibility to monitor, inspect and recommend tourism establishments for licensing. Unlicensed accommodation establishments, even if they offered meals to their guests, were excluded from this study. Apartments, villas and other facilities that did not provide a meal option to their guests were also excluded.

A sample seventy-three (73) licensed hotels and guesthouses or 21.3% of the population were randomly selected. The principal selection criteria included the location of the establishment because the researchers were interested in having a sample representative of establishments in Jamaica's six resort areas—Montego Bay, Ocho Rios, Port Antonio, Kingston, Mandeville and the South Coast and Negril—offer of food service and the size of the establishment. Forty-nine (49) completed responses were received from establishments of varying sizes. Thus, the response rate was approximately 67% of the total sample.

A questionnaire comprising twenty-six (26) questions was developed to aid in the data collection. This included primarily closed-ended questions designed to elicit the source of agricultural produce used in these establishments, the approximate value of the expenditure on agricultural produce, as well as the challenges that were faced in securing agricultural produce from local farmers. A pilot study was first conducted, targeting 10% of the hotels in Kingston resort area between July and August 2016. The information received helped to clarify the reliability and validity of the questionnaire, and helped the researchers redesign the data collection instrument. The data for this chapter was collected between May 2017 and January 2018. The electronic link to the survey instrument was sent out to personnel from the seventy-three (73) establishments selected via email. The personnel contacted included hotel managers, food and beverage managers, and owners who served in multiples roles, especially in guesthouses.

The data collected was analyzed using STATA 16.0. Descriptive statistics are presented in the form of tables and graphs. Given the nature of the data collected, categorical, inferential statistics was primarily done using the Chi-squared goodness-of-fit tests and the two-sample Binomial tests. The Chi-squared test is used to determine if there are differences in the expected and observed frequencies across different categories—as in data collected using Likert scales. The Binomial test is used to determine if there are differences in the proportion of success/failure (yes/no) in the two samples.

Findings

The results of the survey are presented to capture the response of the participants to the research questions based on success criteria that were articulated. This includes an examination of the size of the establishment, their occupancy and purchasing patterns and challenges experienced procuring food items from local sources.

Of the forty-nine (49) respondents, 80% of the establishments contained one hundred (100) rooms or less. Only 14% of the sample had over 300 rooms. 71% of the accommodation establishments enjoyed an average occupancy of over 50% per annum. Only 8% had an occupancy rate of more than 90%, while 10% had an occupancy rate of less than 30% per annum. Most of the establishments did not include a meal as part of their room package.

Consistent with the size of the property, 63% of the respondents included at least one meal while 22% offered the all-inclusive meal plan. Local markets and local farmers were identified as the main source of agricultural produce for 83% of the respondents. More than half of the establishments (57%) sourced produce from local farmers. Of note, only 12% identified the intermediaries or middlemen as their main source of produce, and none of the respondents indicated that imports were the major source for agricultural items.

The data suggests in Table 11.1 that larger properties are more likely to depend on local farmers as their main source of agricultural produce. Surprisingly, the highest rate of utilization of intermediaries as their main source of agricultural produce (39%) was found in the properties with less than 25 rooms. The smallest properties also depended on the local markets for their produce. When asked to identify the agricultural produce purchased by vendor type, local farmers and local markets were the main sources of ground provision, fruits and vegetables, and condiments, while most meat and seafood were sourced from intermediaries or imported.

The data was examined to determine if there are any differences in the type of agricultural produce purchased based on the size of the property and vendor type. Given small sample sizes, the only statistical difference was meats purchased from local farmers. Table 11.2 indicates that properties with less than 25 rooms and those with 101–300 rooms were more likely to purchase meat from local farmers (Observed/Expected > 1). The difference in purchasing patterns of meat from local farmers based on property size was statistically significant at the 5% level.

As shown in Table 11.3, most of the respondents had no major problems securing ground provisions (77%) and condiments (76%). These proportions were statistically different from 50% using the Binomial test, assuming that respondents were equally likely to select yes or no. A

Table 11.1 Major source of agricultural purchases by size of property

Size of establishment	*Local farmers*	*Intermediaries*	*Local market*	*Supermarket*
Less than 25 rooms	39% (7)	17% (3)	39% (7)	6% (1)
25–100 rooms	59% (10)	12% (2)	24% (4)	6% (1)
101–300 rooms	100% (3)			
More than 300 rooms	100% (4)			

The numbers in parentheses represent the number of properties

Table 11.2 Purchase of meats from local farmers by size of property

Size of establishment	*Observed*	*Expected*	*Observed/Expected*
Less than 25 rooms	10	7.7	1.30
25 - 100 rooms	4	6.6	0.61
101 - 300 rooms	3	1.1	2.73
More than 300 rooms	1	2.6	0.38
Pearson χ^2	9.39	p -value =	0.025

higher proportion of respondents had major problems securing fruits and vegetables, 62 and 64%, respectively, while only a small percentage of the respondents found that meats (32%) and seafood (37%) were problematic to obtain. When the size of the properties is considered, the establishments with less than 25 rooms appear to experience the most problems purchasing produce overall. The difference across size of the establishment, however, was not statistically significant.

Additional information on the types of problems that were encounter in purchasing from local farmers is presented in Table 11.4. Consistency in supply was the major problem indicated by 71% of the respondents. Delivery was more likely to be a problem for smaller properties, while the procuring quality produce appears to be more problematic for larger properties.

The respondents expressed agreement with the factors that are considered when planning their menu and purchasing produce. The cost of produce (71%), availability (66%), customer preference (66%) and a preference for local produce (57%) were major considerations. There were no statistical differences in these sentiments based on the size of the property.

Discussion

Several themes have emerged from the data and have been analyzed based on the research questions posed. In addressing research question #1, it was observed that accommodation establishments showed a preference for purchasing their agricultural produce from local sources rather than imports. This is an important finding as this linkage can reduce leakages and strengthen the multiplier effect from tourist expenditures. Specifically, the data suggests that there is a strong preference for purchasing from local farmers since 83% of the respondents procured their agriculture

Table 11.3 Problems purchasing agricultural produce from local farmer

Size of establishment	*Meat*		*Seafood*		*Ground provision*		*Vegetables*		*Fruits*		*Condiments*	
	No	*Yes*	*No*	*Yes*	*No*	*Yes*	*No*	*Yes*	*No*	*Yes*	*No*	*Yes*
Less than 25 rooms	5	5	4	6	8	3	4	7	3	7	7	3
25–100 rooms	6	0	5	1	6	1	5	4	5	4	6	1
101–300 rooms	1	1	2	0	2	1	0	3	1	2	2	0
More than 300 rooms	1	0	1	0	1	0	0	2	1	2	1	1
Total	13	6	12	7	17	5	9	16	9	15	16	5
Total Percent	68%	32%	63%	37%	77%	23%	36%	64%	38%	62%	76%	24%
Binomial Test	p value 0.083		p value 0.179		p value 0.008		p value 0.115		p value 0.154		p value 0.013	

Table 11.4 Problems purchasing agricultural produce from local farmers

Size of establishment	*Consistency*		*Delivery*		*Quality*		*Poor presentation*	
	No	*Yes*	*No*	*Yes*	*No*	*Yes*	*No*	*Yes*
Less than 25 rooms	4	7	6	4	6	5	7	4
25–100 rooms	3	4	6	3	4	4	5	2
101–300 rooms	0	3	1	2	0	3	0	3
More than 300 rooms	0	3	1	0	0	2	1	0
Total	7	17	14	9	10	14	13	9
Total Percent	29%	71%	61%	39%	42%	58%	59%	41%
Binomial Test	p value = 0.032		p value = 0.202		p value = 0.271		p value = 0.262	

produces from local sources. Further, 86% of the respondents expressed that a preference for local foods was an important factor in preparing menus. This preference is also consistent with studies done in other developing countries. For example, in Crete (Andriotis 2002) and Nepal (Shah and Gupta 2000), it was found that most of the hospitality businesses purchased supplies from local sources. Adiyia and Vanneste (2018) found that in Uganda fruits and vegetables were purchased from local sources as opposed to being imported, while in Cancun hotels purchased over 80% of their fruits and vegetables from local sources. The limited shelf life of fruits and vegetables and the increasing demand for the freshest ingredients on menus (Deere and Royce 2019) may account for this finding.

Focusing on research question #2, this study did not find that the accommodation establishments relied on intermediaries or middlemen for their supplies in Jamaica. Interestingly, this is unlike the evidence presented from Indonesia (Kang and Rajagopal 2014), Uganda (Adiyia and Vanneste 2018) and Cancun (Torres and Henshall Momsen 2004). In fact, only 12% of respondents identified middlemen as their main source of agriculture produce. More specifically, intermediaries were utilized more often (39%) by the properties that had less than 100 rooms. They also showed a greater preference for procuring produce from local markets as shown in Table 11.1. It should be noted that intermediaries (middlemen) and to a lesser extent local market may procure the produce eventually sold to these establishments from local farmers as well as from imports.

Of note, the study also found that meats and seafood items purchased by the properties in Jamaica were secured mainly from imports. This is consistent with other studies examined. For example, Adiyia and Vanneste (2018) and Torres (2003) found that meat and other proteins were often imported rather than sourced locally in Uganda and Cancun, respectively. Conversely, the data from Jamaica suggests, however, that smaller properties rely on local farmers for meat. This may indicate that the local agricultural sector still lacks the capacity to produce meat and seafood items in sufficient quantities to meet the demands of the wider tourism industry. The findings that mid-sized properties relied on middlemen (possibly imports) and that meats and seafood were imported point to potential weaknesses in the supply chain in these areas.

When examined by their size, the smaller properties, with less than 25 rooms, obtained agricultural produce from a wider variety of sources, with the major purchases being from local farmers and from local markets. More than half of the properties in the survey showed a preference for local farmers. Further, the larger properties with over 100 rooms identified local farmers as their main source of all produce. Therefore, to answer research question #3, the findings suggest purchasing patterns do vary by property size, depending on the specific item being purchased.

For research question #4, the study found that despite the tendency to purchase most of the fruits and vegetables from local sources, the results indicate that this purchase was most challenging. This may be because of factors such as consistency in supply, the quality of the produce, the quantity provided as well as the presentation or appearance of the food items available for purchase as identified in the literature (Torres and Henshall Momsen 2004). The value chain approach can be used to address these challenges to ensure greater and more successful integration of the operational activities of the local farmers and the properties.

Given the preference and utilization of local sources of agricultural produce, this study indicates that there is a strong relationship between the agriculture and tourism sector. Except for the need to purchase meats and seafood from intermediaries, the study indicates that the local farmers are positioned to supply the needs of the tourism industry. Of note as well is the apparent willingness of the sector to increase production to facilitate import substitution as outlined by the Tourism Demand Study (MTE

2015). This study further highlights the potential for stronger multiplier effects by exploiting the value chain through linkages between these two critical and important sectors, thereby reducing leakages from imports.

Conclusion

The objective of this chapter is to explore the buying patterns of agricultural produce by accommodation establishments in Jamaica and identify challenges in this supply chain. The rationale for this research is to determine what steps can be taken to strengthen the linkage between the agriculture and tourism sectors to build resilience for both sectors and the economy in general, post-COVID-19. The results indicated that the nature of the relationship between local farmers and accommodation establishments (regardless of size) is that of a willingness of these establishments to use local farmers as their primary source of agriculture produce. In order to reinforce the supply chain between the local farmers and the establishments, additional work is required to ascertain the local farmers perspectives, specifically to determine what investments (technology, human, capital) are required to increase the capacity/capabilities of the local farmers to meet the demand from the tourism sector, especially for meats and seafood.

Notwithstanding the willingness of accommodation establishments to purchase from local farmers, there are several challenges—consistency and quality of the produce—within this supply chain that must be addressed. Failure to adequately address these challenges may result in these establishments sourcing agricultural goods from imports. Increasing imports of agricultural goods will increase leakages, resulting in the lowering of the tourism multiplier number, which would adversely affect not just the agricultural sector, but the entire Jamaican economy. Thus, it is imperative that further research is done to determine how leveraging existing institutional infrastructure within RADA, Agri-Linkages Exchange (ALEX) and tourism associations (e.g., Jamaica Hotel Tourist Association) can enable the farmer to understand the needs of the tourism industry. This would foster a better understanding of how their products are being utilized and why specific standards and quality of farm produce are important. Similarly, the procurement staff within the tourism industry will also gain an understand of the farmers' capacity and production cycles. This synergistic relationship will enhance the chances of locally grown produce being incorporated into the menu, thus, securing a stable supply chain for both the properties and the farmers.

Backward and forward linkages have a positive impact on the country's economy. The multiplier effect magnifies the positive effect of linkages to a greater level through several industries. For the tourism industry to remain resilient, the stakeholders involved need to engage in the needed contractual agreements and usage of digital technologies to revitalize and grow linkages with the agricultural sector to ensure the maximum possible economic benefits.

References

Adiyia, B., and D. Vanneste. 2018. Local Tourism Value Chain Linkages as Pro-Poor Tools for Regional Development in Western Uganda. *Development Southern Africa* 35 (2): 210–224.

Anderson, Wineaster. 2018. Linkages Between Tourism and Agriculture for Inclusive Development in Tanzania: A Value Chain Perspective. *Journal of Hospitality and Tourism Insights* 1 (2): 168–184.

Andriotis, K. 2002. Scale of Hospitality Firms and Local Economic Development—Evidence from Crete. *Tourism Management* 23 (4): 333–341.

Belisle, F.J. 1983. Tourism and food production in the Caribbean. *Annals of Tourism Research* 497–513.

Deere, C.D., and Fred Royce. 2019. Provisioning Cuba's Private Restaurants: Linking Tourism and Agriculture. *Tourism Planning & Development* 16 (6): 675–695.

Figueroa, Mark. 2004. W. Arthur Lewis versus the Lewis Model: Agricultural or Industrial Development. *The Manchester School* 736–750.

Hsieh, Hsin-Jung., and Shiann-Far. Kung. 2013. The Linkage Analysis of Environmental Impact of Tourism Industry. *Procedia Environmental Sciences* 17: 658–665.

Jamaica Tourist Board. 2015. Tourism Information Publishing Site (TIPS). *Annual Travel Statistics*. Accessed January 15, 2017. https://jtbonline.org/report-and-statistics.

———. 2019. Tourism Information Publishing Site (TIPS). *Annual Travel Statistics*. Accessed December 16, 2020. https://jtbonline.org/report-and-statistics.

———. 2020. Monthly Statistical Report. October. Accessed December 16, 2020. https://www.jbtonline.otg/report-and-statistics/monthly-statistics.

Kang, S., and L. Rajagopal. 2014. Perceptions of Benefits and Challenges of Purchasing Local Food Among Hotel Industry Makers. *Journal of Foodservice Business Research* 17 (4): 301–322.

Khan, H., Phang, S.Y., and Toh, R.S. 1995. The multiplier effect: Singapore's hospitality industry. *Cornell Hotel and Restaurant Administration Quarterly* 36 (1): 64–69. Research Collection School Of Economics.

Kweka, J., Oliver M., and Adam B. 2003. The Economic Potential of Tourism in Tanzania. *Journal of International Development* 15: 335–351.

Lacher, R. G., and Nepal, S. K. 2010. From leakages to linkages: Local-level strategies for capturing tourism revenue in northern Thailand. *Tourism Geographies* 12 (1): 77–99.

Lewis, W.A. 1983. The Shifting Fortunes of Agriculture. In *Selected Economic Writings of W Arthur Lewis*, ed. M. Gersovitz, 3–34. New York: New York University Press.

Mathouraparsad, S., and A. Maurin. 2017. Measuring the Multiplier Effects of Tourism industry to the Economy. *Advances in Management & Applied Economics* 7 (2): 123–157.

Ministry of Tourism & Entertainment. 2015. *Tourism Demand Study*. October 6. Accessed December 16, 2020. www.mot.gov.jm/page/tourism-demand-study-0.

Mitchell, Johnathan, and Caroline Ashley. 2009. *Value Chain Analysis and Poverty Reduction at Scale: Evidence from Tourism is Shifting Mindsets*. London: Overseas Development Institute.

Mshenga, P.M., and R.B. Richardson. 2013. Micro and Small Enterprise Participation in Tourism in Coastal Kenya. *Small Business Economics* 41: 667–681.

Nguni, Winnie. 2016. Tourist Hotels' Procurement Strategies and Smallholders Upgrading in Local Linkages: Evidence from Horticultural-Tourism Value Chain in Tanzania. *33rd International Business Research Conference*. Dubai: International Business Research Conference.

Ntibanyurwa, A. 2006. Tourism as a Factor of Development. *WIT Transactions on Ecology and the Environment* 97: 73–84.

RADAAdmin. n.d. *Alex—Agri-Linakges Exchange*. Accessed 12 October 2020. https://www.rada.gov.jm/cpntent/blog/alex-agri-linkages-exchange.

Roy, H., C.M. Hall, and P.W. Ballantine. 2017. Trust in Local Food Networks: The Role of Trust Among Tourism Stakeholders and their Impacts in Purchasing Decision. *Journal of Destination Marketing & Management* 6 (4): 309–317.

Shah, K, and V Gupta. 2000. *Tourism, the Poor and Other Stakeholders: Experience in Asia*, Ed. Charlotte Boyd. London: Overseas Development Institute.

Sims, Rebecca. 2008. Food, Place and Authenticity: Local Food and the Sustainable Tourism Experience. *Journal of Sustainable Tourism* 17 (3): 321–336.

Statistical Institute of Jamaica. n.d. Accessed December 16, 2020. https://statinja.gov.jm/NationalAccounting/Annual/NewAnnualGDP.aspx.

Telfer, D.J., and G. Wall. 1996. Linkages between Tourism and Food Production. *Annals of Tourism Research* 635–653.

Torres, R. 2003. Linkages Between Tourism and Agriculture in Quintana Roo, Mexico. *Annals of Tourism Research* 546–566.

Torres, R., J. Henshall, and Momsen. 2004. Challenges and Potential for Linking Tourism and Agriculture to Achieve Pro-Poor Tourism Objectives. *Progress in Development Studies* 4 (4): 294–318.

Torres, Rebecca. 2002. Towards a Better Understanding of Tourism and Agriculture Linkages in the Yucatan: Tourist Food Consumption and Preferences. *Tourism Geographies* 4 (3): 202–306.

Trebicka, Brunela. 2016. Tourism as a Multiplier Effect in Economy: The Case of Albania. *International Journal of Business and Management Invention* 5 (1): 17–21.

Veal, Anthony James. 2018. *Research Methods for Leisure and Tourism*. Harlow: Pearson .

Walker, Brian, C.S. Holling, Stephen Carpenter, and Ann Kinzig. 2004. Resilience, Adaptability and Transformability in Social–Ecological Systems. *Ecology and Society* 9 (2).

Wilson, Dwayne. 2020. Building Tourism Resilience to Thrive in the 'New Normal'. *Ministry of Tourism*. January 7. https://www.mot.gov.jm/speeches/building-tourism-resilience-thrive-%E2%80%9Cnew-normal%E2%80%9D.

CHAPTER 12

Collaborative Tourism Entrepreneurship as a Community Resilience Strategy: A Case Study of Castara, Tobago

Shinelle Smith and Leslie-Ann Jordan

Introduction

The Caribbean region has a deep-rooted history in tourism, being perceived as the most tourism-dependent region in the world (The Green Hotelier 2017). Several small island developing states (SIDS), like Tobago, have forged an economic reliance on the tourism industry to support their development. According to a 2005 World Travel and Tourism Council (WTTC) report on the impact of travel and tourism on jobs and the economy of Trinidad and Tobago, "Travel and Tourism in Tobago is expected to account for 46.0 percent of the island's GDP and 56.8 percent of total island employment. Furthermore, Travel & Tourism visitor exports are projected to account for nearly 96 percent

S. Smith
Tobago Hospitality and Tourism Institute, Blenheim, Trinidad and Tobago

L.-A. Jordan (✉)
University of the West Indies, St. Augustine Campus, Trinidad and Tobago
e-mail: leslie-ann.jordan-miller@sta.uwi.edu

A. Lewis-Cameron et al. (eds.), *Managing Crises in Tourism*,
https://doi.org/10.1007/978-3-030-80238-7_12

of Tobago's total exports" (WTTC 2005, 4). This positions Tobago as a highly dependent tourism-based economy.

However, despite such heavy reliance on the industry, on December 15th, 2016, at a news conference held at the Hyatt Regency hotel in Port-of-Spain, Trinidad, the President of the Tobago Hotel and Tourism Association (THTA), Mr. Christopher James, expressed fears that Tobago's tourism industry will be non-existent in the next few years if something is not done to boost the sector. The COVID-19 pandemic, coupled with restricted international border closure, has since caused arrivals to Tobago to further plummet. According to statistics sourced from T&T's Ministry of Tourism, Culture and the Arts, arrivals fell from 388,576 in 2019 to a mere 78,148 in 2020. As a consequence of this decline, the Tobago accommodation sector has experienced significant challenges in sustaining occupancy levels.

Apart from these extenuating circumstances, the aforementioned WTTC report further articulates limitations based on the internal operations of Tobago's industry, suggesting that in addition to the above challenges "they (tourism businesses) are hampered by an unwillingness to break down competitive instincts and work together for the good of the island's industry as a whole – which will, in turn, provide significant benefits to individual enterprises" (WTTC 2005, 42). In response to such tribulations, hospitality and tourism business owners in Castara, a small fishing village located in Tobago, designed a collaborative, entrepreneurial tourism model to coordinate and promote tourism development, as well as to guard against shocks in the industry. This model was initially developed "pre-pandemic" in 2015, whereby small business owners launched the Castara Tourism Development Association (CTDA) to oversee tourism development initiatives in the community.

At the core of CTDA's framework is the promotion and unison of small businesses to collectively market and develop tourism products within the community in a sustainable fashion. To achieve this, an attempt is made to reduce direct competition with each other and in the same vein, to encourage guests to visit and contribute to other establishments within the CTDA, creating, of sorts, a rural community tourism hub. Mr. Louis Lewis, Chief Executive Officer of the Tobago Tourism Limited, suggests that the model has proven successful in attracting and retaining tourists. In a 2019 article published by LoopTT, Lewis states: "Castara is the best example in the entire Eastern Caribbean of community tourism where a community has done everything that is required, embraced tourism and presented a product that is highly sought after. Hence the reason why their level of occupancy is traditionally counter-cyclical to the visitor flows...there is a constant high occupancy throughout the year upwards of 70 percent and the only other properties that do those things are your top line all-inclusives." It is particularly interesting to note that such success occurred during a period whereby both Trinidad and Tobago were facing declining occupancy levels for a considerable number of larger accommodation units.

For the villagers of Castara, this move to a collective form of tourism development was the result of a declining fishing industry, of which most of the community was actively involved. Once again, given the occurrence of the COVID-19 pandemic, the community is now faced with the need to establish or reinvent its tourism offering to sustain their livelihood. They are in a unique position whereby the product may require tailoring to meet the needs of a new source market, namely domestic or possibly regional visitors as opposed to their returning visitor base, mostly originating from the United Kingdom, Canada or Germany.

This preliminary research took place within Castara during the period March to July 2017 and again between February and April of 2021. It consisted of critical analysis of the entrepreneurs and activities within the community as it relates to tourism-based collaborative entrepreneurship as a community reliance strategy. It also examines whether the collaborative approach can be deemed sustainable utilizing a participatory model which is largely reliant on the promotion of local entrepreneurship. A comparative overview was also considered to review the operational aspects of the model, both before and during the pandemic. Against this background, this chapter will examine some of the main components of this collaborative model, ultimately with the goal of determining whether the model is sustainable and can be replicated in different communities in Tobago to support entrepreneurship and overall diversification within the tourism industry.

Tourism and Entrepreneurship

Curran and Burrows (1986, 269), as cited in Morrison et al. (2009, 8), refer to entrepreneurship as "the creation of an economic enterprise based on a new product or service which differs significantly in the way its production is organised or in its marketing." This is a generally accepted interpretation of the field; however, entrepreneurship itself is a complex and dynamic subject area with varied applications. Apart from an understanding of the societal and intrinsic determinants impacting the entrepreneur, entrepreneurship may be further investigated based on the physical environment in which it functions. While the prevailing trend is the assessment of entrepreneurial clusters which support technological innovation in geographical zones such as Silicon Valley, there is a growing movement of evolving entrepreneurial ecosystems in rural communities. According to Lordkipanidze et al. (2004, 787), "rural development is increasingly associated with entrepreneurship, which is considered a central force of economic growth and development." Many rural communities, like Castara, are now turning to tourism-related enterprises.

Several researchers have investigated entrepreneurial development in a rural context; however, rural communities and their alignment to the tourism industry should gain more consideration (Wortman 1990; Korsgaard et al. 2015). Furthermore, questions have been asked as to the perceived preferences of the new traveller and whether rural communities are likely to attract or repel tourists based on their unique offering. According to Hall (2004), as cited in Polo and Frias (2010, 28), "rural tourism development can act as an agent for the transformation of rural areas, allowing an inflow of resources and liquidity into communities through the expenditure of tourists and the creation of new small businesses and employment." Indeed, "the role of tourism entrepreneurs can be vital for the development of rural areas; therefore, it is essential to find new means of livelihood and alternatives for entrepreneurs" (Lordkipanidze et al. 2004, 791).

Entrepreneurship in tourism has received little attention and according to Ateljevic and Page (2011, 28) "the amount of entrepreneurship research in the leading hospitality and tourism management journals has not increased over the past 21 years." Although this theoretical void exists, it has not hampered the development of entrepreneurship within the tourism industry. In fact, there are new and emerging forms of entrepreneurship, particularly within rural communities and specific to the development of tourism entrepreneurial ecosystems. According to Mason and Brown (2013, 5), an ecosystem, in its generic sense, may be defined as "a set of interconnected entrepreneurial actors, organisations, institutions and processes which formally and informally coalesce to connect, mediate and govern the performance within the local entrepreneurial environment."

The primary earner for tourism entrepreneurs in these rural communities is usually the provision of accommodation. The accommodation sector is a fundamental element of the tourism product and is the largest sub-sectors within the tourism economy. Tourism accommodation is broadly defined as "any facility that regularly or occasionally provides overnight accommodation for tourists" (Medlik 1996). While the resilience of accommodation owners is the focal point of this research, the ecosystem requires a collaborative approach to warrant it as sustainable. As such, entrepreneurship in the tourism sector should encapsulate the relationship among accommodation owners, food and beverage establishments, tour services and other-related businesses, designed to support sustainable tourism activity.

Examining Collaborative Entrepreneurship in Tourism

According to Cooney (2005, 226), "one of the great myths of entrepreneurship has been the notion of the entrepreneur as a lone hero, battling against the storms of economic, government, social and other environmental forces before anchoring in the harbour of success." Carson et al. (1995) also argued that "entrepreneurs do not function in a vacuum, untouched by the social and business systems in which they operate." Indeed, within community settings, specifically in entrepreneurial rural environments, collaborative entrepreneurship is required. Collaborative entrepreneurship is defined as "the possibility of creating something of economic value based on new, jointly generated ideas that emerge from the sharing of information and knowledge" (Miles et al. 2006). In its simplest terms, this form of entrepreneurship supports team work.

As noted by Lyles et al. (1994) as cited in Morrison et al. (2009, 144), new ventures are just as apt to be started by a team of entrepreneurs. Team entrepreneurship is closely related to that of collaborative entrepreneurship. It describes a typology which does not credit entrepreneurial success on account of the sole individual but rather a capability and attitude whereby individual skills are integrated into a group, or team, becoming partners in the business's future evolution. This collective capacity to innovate becomes something greater than the sum of its parts (Morrison et al. 1999, 11). Jones and Haven-Tang (2005, 68) also support an entrepreneurial system which allows for community collaboration, as well as integration with the external tourism enterprises which complement rather than competing with each other. They explained that "partnerships with organisations and businesses that seek to enhance community development initiatives, support fair trade efforts, support responsible tourism and appreciate authentic tourism products are encouraged" (Jones and Haven-Tang 2005, 68).

According to Ateljevic and Page (2011, 24), "Collective entrepreneurship in tourism is not a new practice but one that, in the last decade or so, has been intensified due to strong competition between destinations, changing market trends and above all, to shrinking direct government financial support that has forced the key stakeholders to be more entrepreneurial." According to Hjalager (2010, 12), "entrepreneurship is a crucial factor in the evolutionary redirection of tourism products and increasing competitiveness." This aptly describes the environmental conditions encouraging the upsurge in tourism development in Castara. As a subset of collaborative entrepreneurship, entrepreneurship can also be viewed in the context of a family operation, which is common within rural, tourism communities. Morrison et al. (1999, 12) emphasize that "very often, the small hotel…will involve most of the family performing different roles." Cooney (2005, 228) lists the perceived benefits derived from collaborative entrepreneurship including "pooling financial and physical resources, spreading risk and anxiety, increasing the stock of skills and expertise available and compensating for individual weaknesses."

Community Resilience

According to Abel Duarte (2015), community resilience is defined as the existence, development, and engagement of community resources by community members to survive in an environment characterized by change and uncertainty. Norris et al. (2008, 131) defined it as "[A] process linking a set of networked adaptive capacities to a positive trajectory of functioning and adaptation in constituent populations after a disturbance." And according to Cheer and Lew (2017), community resilience can be described as a form of organization and resilience, and disturbance can range considerably, resulting in great variation in a community's adaptive responses. Therefore, as posited by Norris et al. (2008), community resilience can be intrinsically related to a community's wellness, with regard to quality of life, and the community's functionality.

According to Abel Duarte (2015), entrepreneurship is critical to the restructuring and adaptation of urban and rural economies after an economic shock, because small businesses are particularly adaptable to change. Rural businesses, in particular, increase the diversification of the local economy, hence building wider rural community resilience (Steiner and Markantoni 2013). However, while the perceived benefits of community entrepreneurship in the hospitality industry may present viable alternative to development, Roberts (2008, 576) infers that in the absence of proper regulatory measures and performance indicators, too many inappropriate or poorly operated small-scale developments in the wrong location could proliferate; this could be just as harmful and non-sustainable as a single large development. According to Roberts (2008, 575):

> Despite the numerical dominance of small tourism enterprises in many destinations, little is known about their role in helping destinations progress towards sustainability objectives. The assumption is that, by being small and locally owned, these enterprises automatically contribute to sustainable tourism development.

Furthermore, much of the research surrounding community resilience has mainly focused on the ability of communities to redound after natural disasters such as hurricanes, floods, fires, and earthquakes; man-made disasters such as acts of terrorism and violence; and economic shocks (Bakas 2017; Cheer and Lew 2017; Patel et al. 2017). In summary, a community's resilience is judged by its ability and capacity to come together to work toward a communal objective (Berkes and Ross 2013). This chapter investigates the interplay between collaborative entrepreneurship and community resilience in Castara.

Case Study: Castara Tourism Development Association

The Castara Tourism Development Association (CTDA) was established in 2015 with the aim of promoting responsible tourism and preserving the environment for the enjoyment of both villagers and visitors alike. A report sanctioned by the Tobago House of Assembly (THA) lists the total accommodation units available in Castara at approximately 200 rooms (Poon 2014) with approximately thirty properties registered under the

CTDA. There are several noteworthy initiatives coordinated by members of the CTDA to support collaboration and sustainability among businesses within the community. Rather than competing directly with each other, there is a noted attempt to share resources within the community, which allows for shared benefits from the overall visitor expenditure. Furthermore, activities are developed using a weekly schedule which will allow for most, if not all businesses to benefit from revenue streams. This schedule promotes activities at individual properties to attract tourists on specific nights. Additionally, other businesses are asked to refrain from promotion of independent events to support revenue generation at the agreed upon locations. Some larger properties also opt not to include breakfast packages within their booking options. This in turn allows smaller restaurants to benefit from the sale of food and beverage during this period.

On a broader level, this collaborative approach to tourism development also attempts to incorporate sustainability by addressing the needs of all businesses within the community, even those outside the parameters of the traditional hospitality sector. It is purported that "...all business in Castara is tourism-related. So, whether you sell fruit or fish, rent cars or rooms, make meals or bread, take tours to the rain forest or reef, teach in the school or collect bamboo for Bonfire night, you have a stake in the future of Castara" (Ganase 2018, n.p.). To further add to such sentiments, it is of significance that the community members not only work together, but they primarily own and manage the various tourism products. Steve Felgate of Castara Retreats (one of the founding members of the CTDA) includes on his property's website that "Castara Retreats is run by locals, staffed by locals, and any activity you choose to do there - will be undertaken under the watchful eye of a local." Environmental preservation is also touted as a significant component of the CTDA model. Felgate says "...that tourism must be founded on principles of shared opportunity, to ensure that it sustains the local economy and culture without damage to the environment" (Ganase 2018, n.p.). Furthermore, Taylor adds that "the purpose of CTDA was to share new ideas and to act cohesively. They are already looking at options for alternative energy, wind and solar."

Over the years, the CTDA has implemented a number of community initiatives, such as building concrete garbage bins and speed humps. In 2018, they were awarded a small grant from the United Nations Development Programme (UNDP) to reduce liquid and solid waste (detergents, plastic bags, and non-biodegradable food containers). According to the UNDP, the project was intended to be implemented over 20 months with a total value of USD 77,940, an in-kind contribution of USD 27,216 and a SGP funding request of USD 49,978. The members of the CTDA indicated that they used the money to raise awareness on the environmental issues which directly affected the village. They also ran a summer camp for local children, a day of river, beach and reef activities for all villagers, and several workshops for business owners.

In what may be their most noteworthy accomplishment to date, in 2019, the CTDA launched the "Sustainability Shop" in CasCreole Bar, Big Bay. This initiative was made possible via the aforementioned GEF Small Grants Programme initiative. In collaboration with industry experts, a proposal was submitted which resulted in funding to support sustainable operations, including the opening of a community shop housing saleable, sustainable products. The Shop sells eco-friendly laundry and cleaning products as well as biodegradable food boxes, bags, and straws, and some of these products are made by community members themselves.

The CTDA is mainly funded by member contributions and "was set up to promote Castara as a responsible eco- tourism destination to enable all the people of the village to benefit now and in the future." The Association is represented by tourism-related businesses; however, there is also regular consultation with the Castara Village Council and the Castara Fisher Folk Association before any developmental plan is underway.

Methodology

This research adopted a mixed methods (qualitative and quantitative) approach, firstly to capitalize on the in-depth feedback from semi-structured interviews with key stakeholders and secondly to gather generalized data pertinent to a sample population of entrepreneurs in the Castara community. Qualitative data collection techniques were selected for the first phase of the investigation, which included personal observation of community activities and in-depth interviews with a sample of four (4) institutional representatives and informal interviews with approximately fifteen CTDA business owners within the community. Interviews were initially conducted between May and June 2017. Some key respondents were interviewed again between February and April 2021. Most of the stakeholders responded in line with their original feedback which provided an assurance of reliability. The population for the qualitative research represented only accommodation and restaurant owners in Castara, two subsets of business owners in the community. The research did not consider other community business-related activities such as tour guide services and taxi services as the scope of entrepreneurial interests would have been too large for this investigation.

A sequential approach allowed for sufficient time to analyze the responses collected from these interviews before the second phase of quantitative data collection. Surveys were then disseminated to all accommodation owners in Castara as a means of quantitative data collection. Responses were received from twenty-five (25) stakeholders. The varied methods were implored to compensate for the deficiencies of any singular approach. Data were analyzed using Microsoft Excel and appropriate graphs and charts generated.

Findings and Discussion

Based on the research findings, the researchers identified five (5) main areas which have impacted Castara's collaborative tourism entrepreneurial development: (i) Stakeholder Collaboration, Consultation and Partnerships; (ii) Views on Sustainability; (iii) Institutional Capacity; (iv) Promotion and Marketing; and (v) Training and Development. The following themes are herein discussed in an attempt to enable further investigation and discussion on the collaborative, entrepreneurial model used by the CTDA.

Stakeholder Collaboration, Consultation, and Partnerships

Patel et al. (2017) noted that public involvement and stakeholder support were key elements of community resilience and this included having local participation and community representation in strategic planning at all levels. For example, "...public involvement may involve having local leaders who understand and represent a community's uniqueness and aspirations. A sense of community empowerment can be an additional output of public involvement in governance and leadership" (Patel et al. 2017, 10). As such, it is important to analyze and discuss the CTDA's relationship with some of its main stakeholders, as the sustainability of the CTDA depends in part on partnerships and collaboration with other NGOs, as well as public sector organizations.

The Tobago House of Assembly (THA) and by extension, the Division of Tourism, Culture and Transportation is a valuable stakeholder lending support to the local industry. On their website, their responsibilities include "establishing, standardizing and sustaining the island's tourism product in a manner consistent with the repositioning strategy for Tobago as a tourist destination." While evidence exists of their efforts, their interactions with the CTDA have been somewhat problematic. Most respondents (21 out of 25) were dissatisfied with the vision of the THA regarding tourism development and the lack of consultation afforded to the community. In fact, it is believed that these were the main factors which initially supported the formation of the CTDA (Fig. 12.1).

Notwithstanding such views, the THA has recognized Castara's contribution to the tourism industry. To this end, a report was commissioned in 2014 to provide recommendations on the way forward to Castara and by extension, other communities with similar characteristics. Regrettably, although most of the business owners were actively involved in supporting the research, the CTDA indicated that they were not provided with the final report for review. As such, it was believed that the analysis primarily supported data collection and did not offer reommendations to their community, which could have, in turn, offered further insight into best practice.

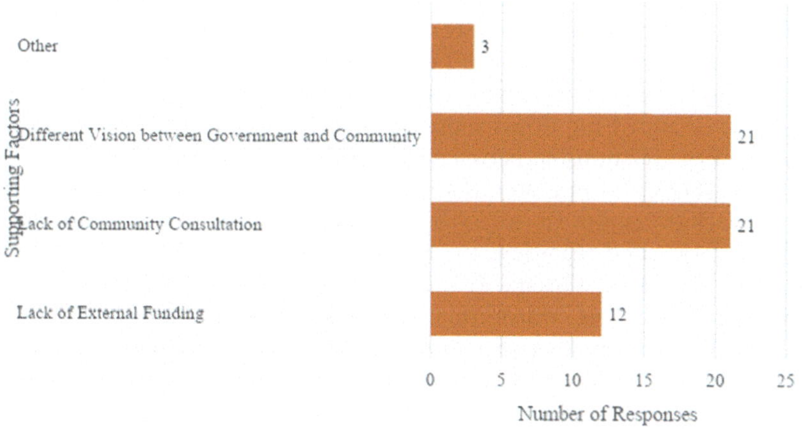

Fig. 12.1 Factors promoting the formation on the CTDA

Another arm of the THA is the Community Development Enterprise Development and Labour Division (Community Division) which aims to "improve the quality of life of Tobagonians through community mobilization, while preserving, promoting and appreciating our unique cultural traditions in the global environment." This Division aims to support entrepreneurship and community development and in doing so, refurbished a Community Centre in Castara in 2018. The Enterprise Division also provides an incubator program for entrepreneurs, though as of 2017, no tourism-based company had yet benefitted from such. Mr. Eastman (Entrepreneur) is of the opinion that the Centre is underutilized and can function more efficiently as a training facility for introductory courses in business management and customer service for entrepreneurs in Castara. However, despite the positive contributions of the Community Division, their relationship with the CTDA deteriorated when the Division established a Fishing Depot in Castara which was met with great resistance some of the community members. Most of the small business owners of the CTDA lamented that the construction of the facility was directly on the beach which obscured views for some guest houses—a major selling point for tourists. Mr. George (Entrepreneur) indicated that the CTDA pleaded with the Division to reverse the decision based on two main accounts. Firstly, fishing was no longer the main income earner for the community, so it was believed that resources ought to have been directed toward tourism-related initiatives. Secondly, the choice of materials used for the construction of the facility did not appear eco-friendly which was a direct contradiction to the brand of sustainability within the community. Despite such sentiments, the construction proceeded. Such interactions and heavy-handed top-down approaches to development have negatively impacted the relationship between the CTDA and THA.

The CTDA also indicated that they grew weary of the government's overall lack of responsiveness to some of the community's needs. Another situation was cited regarding the lack of a functioning Automatic Teller Machine (ATM) in Castara. If tourists were unable to withdraw funds, this limited possible visitor spending within the community. It was compounded by the fact that many of the smaller properties operated on a cash-only facility and would not be in a position to capitalize on spending via credit cards or mobile applications. Furthermore, the CTDA stated that the involvement of the THA in community events is not readily supported by all villagers as governmental involvement may have the potential to detract from an authentic experience. Reference was made to the "Tobago Blue Food Festival," a community staged event, in celebration of the indigenous dasheen crop, a versatile and edible form of produce grown in Trinidad and Tobago. Interviewees, which included members of the CTDA, indicated that the event lost some of its appeal and is now managed by the THA with little community involvement. It should be noted that the THA disagreed with this assessment and the then Secretary of Tourism and Transportation, Assemblyman Tracy Davidson-Celestine stated that "the Blue Food event marks a significant achievement of community and state partnership...which stimulates economic and entrepreneurial activity." Such competing ideologies suggest that a collaborative approach to community-based events may be required to satisfy the needs of the various stakeholders.

As it relates to the governance structure in place to support the CTDA, most of the respondents (92%) believed that the CTDA functions best when controlled by the community. Two respondents (8%) believed that there is room for a more collaborative approach with the THA and community control. With such strong views regarding community management, the role of the THA and the challenges with existing stakeholder collaboration, there is a need for further review of these underlying complexities.

Views on Sustainability

The success of any collaborative model depends on participants having shared views. Therefore, respondents and interviewees were asked about their perspectives regarding authenticity, which is a core value of the CTDA model and a component of socio-cultural sustainability. All interviewees supported the notion that preservation of culture and traditions was vital to the success of the CTDA model and 23 out of 25 respondents rated "local culture" as extremely important. In addition to assessing their outlook on culture, and in keeping with the four (4) main tenets of sustainable tourism development, CTDA's views on the environment were also taken into account. Twenty respondents rated the "preservation of the environment" as extremely important. This is in keeping with the views espoused by the CTDA, which brands Castara as an eco-tourism destination and as such, environmental preservation is deemed a priority. However, while there is more than sufficient evidence of eco-based practices within the community, there may be need for the incorporation of additional interpretive measures to communicate such practices to guests. Although most interviewees reflected views which were in support of environmental preservation, there was no common philosophy communicated to guests which encouraged such practices (e.g., code of conduct, use of air-conditioning, etc.). As such, interpretation seemed to be lacking in this regard. Additionally, there is no institutional policy governing carrying capacity at eco-based attractions within the community of Castara. This could contribute to the deterioration of natural sites should arrivals rapidly increase in an unmanned environment.

Another contentious issue raised involved the competing use of the community's resources. Although members of the CTDA collectively decided on guest activities via a scheduled listing of weekly "things-to-do," one interviewee raised concerns that "some members (may) have a different vision or cater to a different market which results in activities that upset other guests." This may include properties which prefer to cater to nightlife and entertainment which in turn can impact on the satisfaction levels of guests who visit the community for a sense of tranquility and relaxation. These competing environments may offer an opportunity to target a wider demographic; however, it will not reflect a common community philosophy, which the CTDA is in fact attempting to promote.

On a positive note, the community is currently on a recycling drive where all businesses are asked to recycle aluminum and plastic. The CTDA has also partnered with Carib Glass to install recycling bins throughout the community. Councillor Kwesi Des Vignes, in his capacity as Secretary of the Division of Infrastructure, Quarries, and the Environment, THA, "praised the work of the CTDA, noting initiatives like moving toward a plastics-free village." He said that in the village of Castara, community members had adopted a "culture of conservation and respect for the environment." Councillor DesVignes then engaged a private–public partnership which lasted for one year and involved the DIQE, the Division of Health, Wellness and Family Development and Recycling Waste and Logistics Ltd, based in Trinidad. Stemming from their first environment partnership Conference in 2019, a recycling and an education program began in the Castara to sensitize the community about separating their plastics and aluminum.

Notwithstanding these admirable initiatives, there were additional constraints noted in the exercise of conservation and preservation measures. Mr. Taylor, owner and operator of AliBaba's Sea Breeze and Tours and President of the CTDA, also expressed the need for a more integrated approach regarding beach and waterfall clean-ups with more attention needed on developing holistic and sustainable practices. There were also conflicting views among community members managing the transition from fishing to tourism. For example, spear-fishing supports some of the fisherfolk but at the same time, depletes the population at the reef. This activity therefore is likely to reduce the snorkeling experience for visiting tourists, and the reef in itself is one of Castara's most sought after natural attractions. Overall, only two (2) respondents were not particularly driven to support environmental preservation as a priority.

While environmental considerations seemed to have the general support from the community, socio-economic considerations appeared less of a concern. While eighteen members viewed employability via tourism as "extremely important", six (6) respondents did not believe that equitable distribution of wealth was significant (Fig. 12.2). This was an interesting finding as much of the literature on collaborative entrepreneurship discussed "pooling financial resources" as a requirement of similar models. While the community members seemed open to the prospect of job creation, some were less than willing to contribute to equal sharing of financial profits. This could be on account of varying levels of support, available resources and deeper sentiments regarding imbalances within the community hierarchal structure. Collectively, this may have wider implications for the future growth of the sustainable operations within the community as the sustainable model should rely on all four (4) criteria presented in Fig. 12.2.

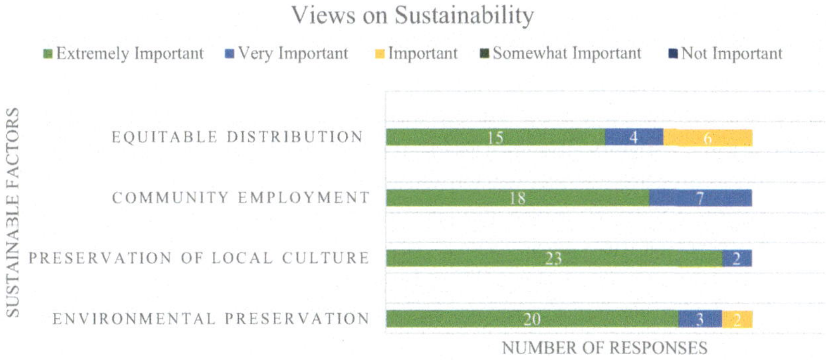

Fig. 12.2 Views on sustainability

Institutional Capacity for Sustained Growth

The existence of NGOs like the CTDA is important to supporting community resilience. As Erkus-Ozturk (2011, 1736) noted, "Newly emerging non-governmental organizations (NGOs) and associations, most of which increase institutional capacity, have become important in stimulating linkages between companies, and thereby increasing the competitiveness of a region. In addition, new institutional set-ups, such as NGOs and trade associations, have become important in tourism regions due to the stimulation they provide to the development of collaborations and joint projects with related actors, leading to a sustainable future in tourism development." Against this background, members of the CTDA were asked to list the perceived weaknesses of the Association. The owner of one of the smaller lodging properties expressed concerns that "larger and older businesses would eventually gain the most out of tourism development." Possibly, as the number of tourism enterprises grow and develop within the community of Castara, equitable distribution of resources may be problematic in the future. Similar concerns were also expressed by the owner of a small restaurant and a larger accommodation property, respectively. In 2017, one stakeholder stated that "if new businesses are established, it will be difficult given the existing structure of the CTDA to promote all tourism (activity equally)" (i.e., with no competing activities among entrepreneurs). Another accommodation owner expressed that "we have to ensure that profits are shared among the community."

The respondents were also asked to list the perceived strengths of the CTDA. More than one member of the Association indicated that "the community is able to coordinate efforts without 'red tape'." Another interviewee stated that "we can supply a good destination which is unlike any other… not what the government believes tourism is about." Such responses support the notion that governmental involvement is not perceived as an entirely necessary inclusion in the operation of the CTDA. Interestingly, in 2020, when governmental funding was made available to support restoration work for accommodation units, several business owners and members of the CTDA benefited from this via the Tourism Accommodation Relief Grant. It would appear that government intervention regarding financial support may be a meaningful inclusion to the development of businesses in CTDA, despite such resounding calls to work apart from formal political forces.

Overall, most of the respondents advocated their support regarding the utilization of a collaborative entrepreneurial structure as a tool for community resilience and development expressing sentiments such as "business owners benefit from their combined strengths," and "there is the ability to offer a unique product supporting community enterprises."

Promotion and Marketing

Although there has been some measure of success regarding the collaborative entrepreneurial model, hindrances were also identified in their marketing and promotional thrust. One of the largest obstacles identified was insufficient marketing with interviewees indicating that there may be a mismatch regarding the marketing channels used and the intended audience (Fig. 12.3).

Figure 12.4 illustrates that the predominant market for businesses in Castara is UK based, capturing 48% of market share. The second largest category was Trinidadians (32%) which suggests further opportunities for domestic tourism. The Ministry of Tourism's "*Stay to Get Away*"

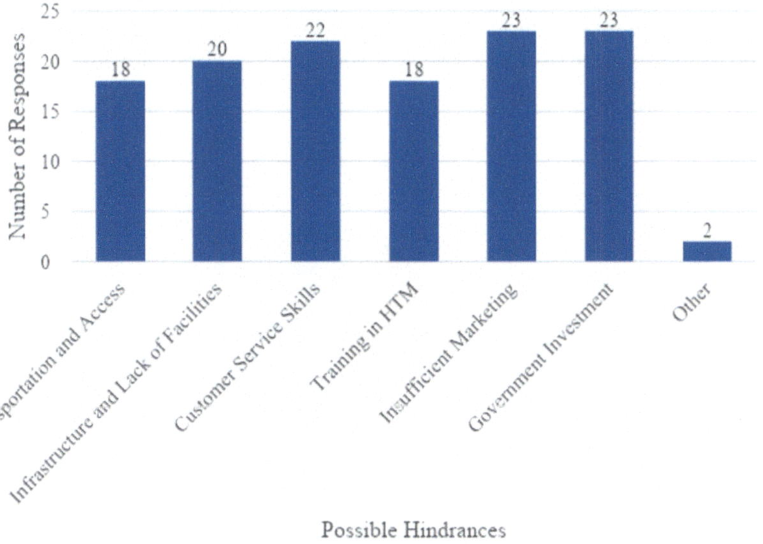

Fig. 12.3 Factors hindering entrepreneurial tourism development

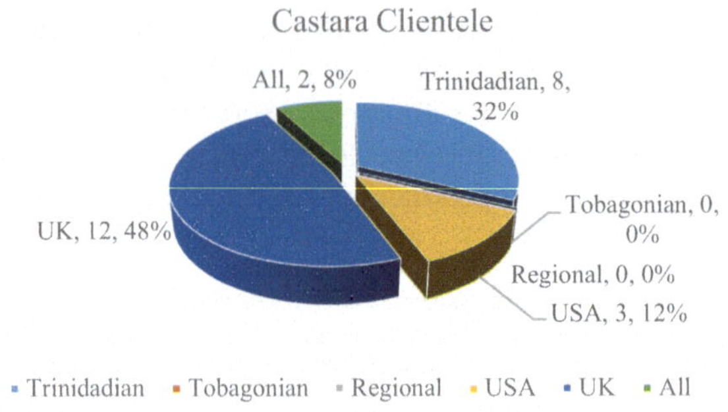

Fig. 12.4 Main clientele for Castara

campaign should explore these findings as the campaign is "part of a more holistic promotional approach to develop community clusters… events and festivals" (Ministry of Tourism 2017). However, upon review of the then campaign details, no Castara businesses were listed as available offerings.

Figure 12.5 reflects that most respondents selected multiple marketing strategies which suggest that marketing efforts are widespread. The leading strategies included independent websites (23 of 25 respondents) and hosted travel sites (17 of 25 respondents). These figures were very

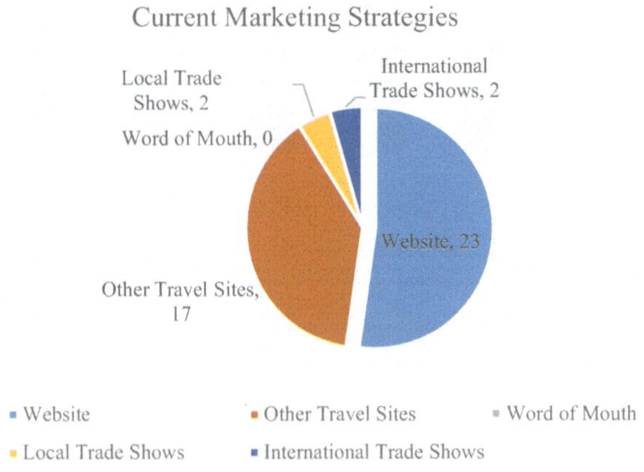

Fig. 12.5 Marketing strategies used

Fig. 12.6 Effectiveness of marketing tools

optimistic for a rural community where members had little to no formal training in the field of marketing.

Despite an active approach to marketing, Fig. 12.6 illustrates that most of the respondents evaluated the success of the marketing strategies from an uninformed perspective. Only 4 of 25 respondents (16%) utilized customer feedback and evaluation in a formal manner. While no formal data capture systems exist, most interview respondents indicated that Expedia and Trip Advisor were largely credited with attracting visitors.

Despite the current efforts by the CTDA to build a website for the promotion of businesses within the community, 12% of respondents were not in favor of this approach and 12% were uncertain that it would yield benefits. This is an interesting finding as the very nature of the CTDA is aimed at supporting a common platform for business and community promotion. One respondent expressed concerns that some businesses would receive far greater levels of publicity than others. This coincides with sentiments expressed earlier regarding concerns about equitable distribution.

The THTA has promoted Tobago's tourism at international trade shows, usually funded by private investment and funding from state entities such as the Tourism Development Company (TDC). The success of these initiatives is aligned to increased occupancy rates at hotels for the following quarter or commitments from tour operators to book bulk packages. The CTDA interviewees did not believe that this approach was beneficial to smaller properties granted that they were not showcased as viable accommodation options. However, Mr. Taylor indicated that although Castara's average accommodation unit provided less than four rooms, the intent of the CTDA is to market the entire community as a 150 plus-room beach resort with varied accommodation options and price points.

As such, according to Mr. Taylor, the business owners in Castara are developing an integrative website which will be used to market the community. This website will draw on the strength that each property will be marketed as one component of the overall product. It will list all community activities, restaurant options, lodging options as a singular package. Despite this ambitious undertaking, a member of the CTDA indicated that technical expertise was lacking and so it may be challenging to provide relevant and updated information for a single website managing a plethora of businesses.

Entrepreneurial Training and Development

According to Suryana (2014) in Salipi and Nugroho (2019), "Entrepreneurship is the essence of the economy of a nation. The entrepreneurial spirit is essential to create better conditions for individuals or groups of people that should be able to contribute a positive value to the wider community by encouraging others and stimulating the national economy." At the heart of this is empowerment as a tool to promote sustainability. The research findings have identified a few areas relevant to training and development that should be supported in order to strengthen the sustainability of the model.

The majority of business owners did not possess a formal business plan (92%). The researchers observed that most of the smaller operations followed successful patterns of larger properties without an understanding of why certain practices were adopted. However, in the absence of a business plan, the CTDA meets frequently to propose short-term developmental initiatives which involve all businesses. Additionally, most respondents were also unsure of their future development plans regarding business operations. However, the larger property owners were interested in developing new facilities.

In terms of education, according to Fig. 12.7, the majority of respondents obtained up to a secondary level education (60%). This finding supports the design and nature of training programs accessible to the community.

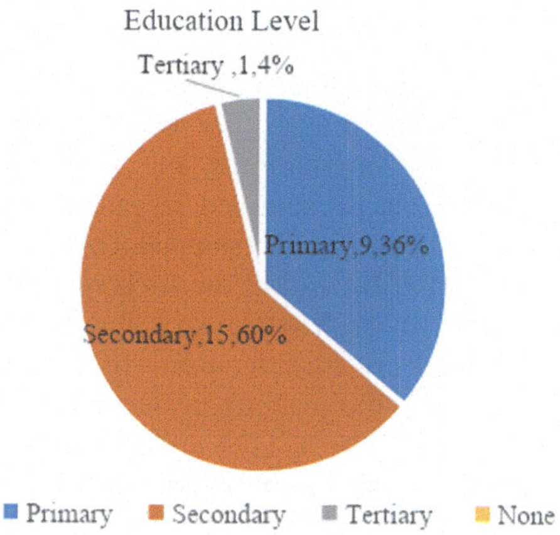

Fig. 12.7 Education levels among entrepreneurs in Castara

Additionally, despite the success of business operations, only 1 of 25 respondents received training in entrepreneurship, while 96% did not. While training is not an absolute requirement for business success, it may prove advantageous for the CTDA to afford some degree of training linked to marketing, accounting, and hospitality law in the context of business ownership. Surprisingly, 76% of respondents did not have any prior work experience in the field of Hospitality and Tourism Management. So, with no direct experience in the industry, no academic expertise in the field and little support from external bodies, it may be challenging for long-term development. Also, these dynamics can affect what Dahles and Susilowati (2015) describe as business or enterprise resilience, which is defined as the "capacity for an enterprise to survive, adapt, and grow in the face of turbulent change."

The CTDA expressed a willingness to receive such training; however, they cited two major concerns. First, as business owners who are actively involved in most, if not all aspects of operations, it would be difficult to undertake a program offered on a full-time basis. Secondly, the preference would be for training opportunities to be made available in Castara (using the Community Centre), rather than encouraging business owners to attend classes in the city which is approximately 45 min away. Some respondents expressed a willingness to receive training virtually as a viable alternative for future professional development.

Conclusion and Recommendations

A proliferation of small apartments and guest houses in Castara, largely operated by families, has been a testament to a marked increase in demand by international and local visitors. The business community of Castara has responded to such development and formed a unique model of entrepreneurship which is supportive of the collective enterprise. Notwithstanding the success of tourism in the Castara community by means of high occupancy, the overall developmental approach must be taken into consideration. In reviewing respondents' views toward community employment, preservation of local culture, and environmental preservation, the CTDA can be regarded as an example of achieving sustainability through collaborative efforts. The only exception seems to be the need for a common philosophy regarding equitable distribution, which in turn would suggest a greater need for equitable collaboration.

With the majority of business owners primarily motivated by profit generation, there may be negative implications for sustainable development should socio-economic considerations surpass the need for socio-cultural and environmental preservation. While the CTDA currently facilitates initiatives in this regard, with most individual business owners void of formal education or training in the field, there is a likely challenge to communicate a vision that is truly shared or understood by all. Findings justify this with some members listing conservation and preservation efforts as a lower priority, despite highlighting the beaches as the main attraction in the community. It creates a paradox of sorts where there is greater need to foster holistic communication about sustainable operations, which will directly affect the tourism product, and by extension, revenue generation.

The hesitance of the CTDA to work with the public sector is also a unique finding as most communities of this nature are heavily reliant on governmental support and financing. Such frameworks are ultimately required to streamline the nature of development alongside other industries. The main source of friction between the CTDA and the THA seems to be a distrust regarding the degree of control afforded to the CTDA by the THA. Additionally, the imposition of THA on community events, though successful, seems to somewhat exclude the views of the community. The preservation of cultural and social norms is a requirement for sustainability, as such planned initiatives which are likely to impact on a community's traditions and way of life requires community buy-in. To support collaborative efforts, the THA can host community meetings to highlight the success stories of joint initiatives with other Tobago communities in an attempt to support public–private sector partnership. However, with the CTDA's belief that there is inadequate consultation and somewhat of an opposing vision between government and community, it is unlikely that successful collaboration will occur under the current approach. A more bottom-up view and inclusive reach are needed to support viable institutional linkages. Additionally, wider sectoral support will be influential in supporting entrepreneurship and empowerment among such marginalized communities, particularly during times of unexpected and unprecedented disruptions within the industry.

The lack of a legal and administrative framework can also present challenges to the long-term sustainability of the CTDA. According to Murphy and Murphy (2004), the government perceives similar associations as disorganized. As such, the absence of clear regulatory guidelines and formal structures for the functioning of the CTDA may support such claims. Additionally, should the intention of the CTDA extend to seek funding and collaborate on wider, sectoral marketing initiatives, legal compliance for all business owners may be required to satisfy the demands of investors to be eligible for support. Creation and adoption of formal systems by the CTDA will allow for a greater degree of legitimacy among participants in this regard.

For the CTDA, the need for capacity building is also needed. According to views expressed earlier by Lerner and Haber (2001, 77), "a lack of managerial skills is one of the main barriers to a venture's success." Furthermore, one respondent agreed that a major challenge is the void in membership as it relates to professional backgrounds in tourism. The CTDA can seek linkages with the Caribbean Tourism Organization to explore the provision of an annual scholarship in the field of Hospitality or Tourism, in an attempt to develop local expertise. Additionally, training may be required in Computer Literacy and the use of new software to assist businesses with managing data entry and forecasting trends. With the shift to a more inclusive, virtual environment for learning, it is also possible to use this season to liaise with community members, develop relevant training programs to meet their needs, and train members of the CTDA how to interact with learning management platforms to increase their competencies.

Managing the size of the market is also a crucial concern in addressing efficiency. The ability to gather collective data on visitors will inform development in Castara and guide the nature of activities, businesses, and appropriate marketing initiatives. Benchmarking against the Baseline Survey conducted by the THA in 2018, which attempted to collect and analyze relevant and timely data in the areas of tourism businesses, primary tourism employment, skills and training and tourism revenue, a similar initiative can be coordinated by the CTDA in conjunction with THA to acquire relevant statistics within the community hub. Investment in a Tourism Centre in Castara as a community-owned asset (similar to the Sustainability Shop) will provide additional employment, be a point of contact for tourist information and serve to collect data regarding visitor trends. Behavioral codes for environment and community sustainability and carrying capacity limitations can also be communicated to visitors through this medium thus further increasing opportunities for interpretation.

The incorporation of innovation in an entrepreneurial environment is deemed a necessity for survival. Training, specific to entrepreneurship education, will be vital to support creativity and modernization of businesses. Provision of rotational short training courses in Hospitality and Tourism Management, Accounting, and Business Planning can be offered to business owners and the wider community. To address the perceived challenges regarding training, and as highlighted by respondents, the Community Centre can be utilized to facilitate such interaction. The THA can also liaise with the THTA in the creation of an incubator program specific to the needs of tourism business owners. As this study is geared toward supporting the industry as a whole, the transfer of success stories, as well as challenges for the CTDA, ought to be communicated to other communities as learning lessons. Facilitation of seminars and workshops on the benefits of rural community tourism will showcase CTDA business owners to the wider tourism community as a means of sensitization.

Overall, a policy framework, unique to tourism entrepreneurship, is required and should include representatives from CTDA or the respective Village Councils where development is intended to take place. While there should be continued outsourcing of support from private entities to offset funding for community outreach activities, collaboration with the government offers a sustainable approach and consistent vision with larger national objectives. Ultimately, while there is room for growth in any model, the CTDA successfully manages to promote a sense of resilience and affords a viable alternative for sustainable entrepreneurial development of rural tourism enterprises in Tobago.

References

Abel Duarte, A. 2015. Resilience in the Context of Two Traditional Spanish Rural Sectors: An Exploratory Study. *Journal of Enterprising Communities: People and Places in the Global Economy* 9 (2): 182–203.

Ateljevic, J., and S.J. Page. 2011. *Tourism and Entrepreneurship: International Perspectives*. London: Routledge.

Bakas, F.E. 2017. Community Resilience through Entrepreneurship: The Role of Gender. *Journal of Enterprising Communities: People and Places in the Global Economy* 11 (1): 61–77. https://doi.org/10.1108/JEC-01-2015-0008.

Berkes, F. and H. Ross. 2013. Community Resilience: Toward an Integrated Approach. *Society & Natural Resources, 26*(1): 5–20.

Carson, D., S. Cromie, and P. McGowan. 1995. *Marketing and Entrepreneurship in SMEs: An Innovative Approach*. New Jersey: Prentice Hall.

Cheer, J.M., and A. Lew. 2017. Understanding Tourism Resilience: Adapting to social, political, and economic change. In T*ourism, Resilience, and Sustainability: Adapting to Social, Political and Economic*, Change, 3–17. Routledge

Cooney, T. 2005. What is an Entrepreneurial Team? *International Small Business Journal* 23: 226–235.

Curran, J., & R. Burrows. 1986. The sociology of petit capitalism: A trend report. *Sociology, 20*(2): 265–279. https://doi.org/10.1177/003803858602 0002007.

Dahles, H., and T.P. Susilowati. 2015. Business Resilience in Times of Growth and Crisis.*Annals of Tourism Research* 51: 34–50.

DesVignes: Think about the Future. 2019, June 6. *Trinidad and Tobago Newsday*.

Erkuș-OZturk, H. 2011. Emerging Importance of Institutional Capacity for the Growth of Tourism Clusters: The Case of Antalya. *European Planning Studies*, 1735–1753.

Ganase, P. 2018. Castara: The Village that Markets Itself. *Contact Magazine*. MEP Caribbean Publishers, https://www.meppublishers.com/castara-the-village-that-markets-itself/#ixzz6sufejK7O.

Hall, D. 2004. Rural tourism development in south-eastern Europe: Transition and search for sustainability. In *The International Journal of Tourism Research, 6*(3): 165–176.

Hjalager, A.-M. 2010. A Review of Innovation Research in Tourism. *Tourism Management*, 1–12.

Jones, E., and C. Haven-Tang. 2005. *Tourism SMEs, Service Quality and Destination Competitiveness*. Oxfordshire: CABI Publishing.

Korsgaard, S., S. Muller, and H.W. Tanvig. 2015. International Journal of Entrepreneurial Behavior & Research. *International Journal of Entrepreneurial Behavior & Research*, 5–26.

Lerner, M., and S. Haber. 2001. Performance Factors of Small Tourism Ventures: The Interface of Tourism, Entrepreneurship and the Environment. *Journal of Business Venturing* 16: 77–100.

Lordkipanidze, M., H. Brezet, and M. Backman. 2004. The Entrepreneurship Factor in Sustainable Tourism Development. *Journal of Cleaner Production* 787.

Lyles, M.A., N.M. Carter and I.S. Baird. 1994. Partnering in establishing new ventures: the experience of Hungary. In *Frontiers of Entrepreneurship Research, Proceedings of the 1994 Babson Entrepreneurship Research Conference*, Wellesley, MA.

Mason, C., and R. Brown. 2013. *Entrepreneurial Ecosystems and Growth Oriented Entrepreneurship*. Hague: OECD.

Medlik, S. 1996. *Dictionary of Travel, Tourism and Hospitality*, 2nd ed. Oxford: Butterworth-Heinemann.

Miles, R., G. Miles, and C. Snow. 2006. Collaborative Entrepreneurship: A business model for continuous improvement. *Organizational Dynamics*, 1–11.

Ministry of Tourism. 2017, June 29. Tourism Baseline Survey launched in Tobago. *Trinidad and Tobago Newsday*. Retrieved from Tourism Baseline Survey launched in Tobago.

Morrison, A., M. Rimmington, and C. Williams. 2009. *Entrepreneurship in the Hospitality, Tourism and Leisure Industries*. Oxford: Butterworth-Heinemann.

Morrison, A., M. Rimmington, and C. Williams. 1999. *Entrepreneurship in the Hospitality, Tourism and Leisure Industries*. Oxford: Butterworth-Heinemann.

Murphy, P., and A. Murphy. 2004. *Strategic Management for Tourism Communities*. Great Britain: Cromwell Press.

Norris, F.H., S.P. Stevens, B. Pfefferbaum, k.F. Wyche, and R.L. Pfefferbaum. 2008. Community Resilience as a Metaphor, Theory, Set of Capacities, and Strategy for Disaster Readiness. *American Journal of Community Psychology* 41 (1–2): 127–150.

Patel, S.S., M.B. Rogers, R. Amlôt, and G.J. Rubin. 2017. What Do We Mean by 'Community Resilience'? A Systematic Literature Review of How It Is Defined in the Literature. *PLOS Currents Disasters*. Feb 1. Edition 1. https://doi.org/10.1371/currents.dis.db775aff25efc5ac4f0660ad9c9f7db2.

Polo, A.I., and D. Frias. 2010. Collective Strategies for Rural Tourism: The Experience of Networks in Spain. *Journal of Tourism Consumption and Practice* 2: 25–44.

Roberts, S. 2008. Sustainability Indicators for Small Tourism Enterprises—An Exploratory Perspective. *Journal of Sustainable Tourism*, 575–594.

Salipi, C., and F. Nugroho. 2019. Resilience of Tourism Entrepreneurship in the Disaster-Prone Area III of Mount Merapi-Indonesia through Social Development Approach. *Conference: Proceedings of the Third Workshop on Multidisciplinary and Its Applications*, 11–14. Medan: WMA.

Steiner, A., and M. Markantoni. 2013. Unpacking Community Resilience through Capacity for Change. *Community Development Journal* 49 (3): 407–425.

The Green Hotelier. 2017, June 24. *The Caribbean, the Most Tourism-Dependent Region in the World*. Retrieved from The Green Hotelier http://www.greenhotelier.org/destinations/the-caribbean/.

Tobago House of Assembly. 2017, June 29. *Tobago gearing up for Blue Food Festival*. Retrieved from Tobago House of Assembly http://www.tha.gov.tt/news/tobago-gearing-blue-food-festival/.

Wortman, M.S. 1990. Rural entrepreneurship research: An integration into the entrepreneurship field. *Agribusiness: An International Journal*, 329–344.

WTTC. 2005. *Trinidad and Tobago: The Impact of Travel & Tourism on Jobs and the Economy*. World Travel & Tourism Council.

PART V

Contemporary Landscape & Reflections

CHAPTER 13

COVID-19 Policy Responses—The Tension Between Lives and Livelihoods in Tourism-Dependent Caribbean Territories

Acolla Lewis-Cameron and Sherma Roberts

Introduction

At the time of writing, the global tourism industry is engulfed in a battle for survival as the industry navigates the fallout from the deadly COVID-19 pandemic. On record as the most severe crisis to hit the tourism industry, the negative impacts of the COVID-19 pandemic are more pronounced on the vulnerable island states in the Caribbean. As discussed in chapter two, island states possess a narrow resource base due to small, dispersed land area, resulting in lower economies of scale (UN-OHRLLS 2011); are remotely located from key markets leading to 'high production and trading costs, limiting investment, competitiveness and the scope for

A. Lewis-Cameron (✉)
University of the West Indies (UWI), St. Augustine, Trinidad
e-mail: acolla.lewis-cameron@sta.uwi.edu

S. Roberts
University of the West Indies, Cave Hill Campus, Barbados
e-mail: sherma.roberts@cavehill.uwi.edu

A. Lewis-Cameron et al. (eds.), *Managing Crises in Tourism*,
https://doi.org/10.1007/978-3-030-80238-7_13

integrating global value chains' (OECD 2018); have high dependence on foreign aid/intervention (Everest-Phillips 2014) and high levels of debt and lack of human capital (UNEP 2014). Consequently, there is a high dependence on a narrow, undiversified production base built primarily on the services sector, particularly tourism and financial services with exportation of few goods (OECD 2018). This places them in a vulnerable state which according to Birkmann et al. (2013, 195) is 'the propensity of exposed elements such as physical or capital assets, as well as human beings and their livelihoods, to experience harm and suffer damage and loss when impacted by single or compound hazard events.'

Caribbean island states are primarily tourism-dependent states that are highly susceptible to external shocks. As such, it is not surprising that these island states have encountered much hardship due to the pandemic. Islands such as the Bahamas, the British Virgin Islands and Barbados whose average GDP contribution and export earnings from tourism exceed more than 55% will experience a 12% plus decline in GDP in 2020 (Coke-Hamilton 2020; Ramrattan 2020). Travel and Tourism job loss at the end of 2020 in the Caribbean region can increase up to 1.9 million while in destinations such as the Bahamas and St. Lucia, as much as 20% of the labor force could be adversely affected (Inter-American Development Bank 2020). According to Meddeb (2020), 'many are now facing debt to GDP ratios over the thresholds for sustainable debt defined by the International Monetary Foundation' (para. 2).

For these tourism-dependent states, the fundamental challenge is the tension between protecting lives and livelihoods. The COVID-19 pandemic reveals important lessons that can be learned regarding the resilience of their tourism systems. It is with this in mind that this chapter employs document analysis to comparatively analyze the responses to the pandemic of six (6) tourism-dependent islands, namely, Barbados, British Virgin Islands, St. Kitts and Nevis, The Bahamas, Antigua & Barbuda and St. Lucia. These island states all exhibit on average a high dependence on tourism as characterized by GDP contribution and export earnings from tourism. The chapter will further propose destination resilience considerations for tourism-dependent islands in this new operating environment.

COVID-19 Tourism Impacts

The stark reality that 2020 would be a rough year for the global tourism industry culminated as the World Health Organization (WHO) declared COVID-19 a pandemic on March 11, 2020. This announcement preceded state of emergency declarations in many US states, bans on large gatherings in countries such as France, suspension of flights from China and Italy and a rising death toll from the pandemic of thirty-four hundred plus. The announcement coupled with a sharp rise in COVID-19 cases and deaths sent the global tourism industry into a tailspin with 100% of destinations introducing travel restrictions including border closures resulting in mass cancellation of travel plans and bookings. For the period January–October 2020, the UNWTO (2020a) indicates that due to travel restrictions, slow virus containment and low traveler confidence, international arrivals fell by 72%. Further, the Asia Pacific region experienced the worse decline of 82% for the same period, followed by the Middle East by 73%, Africa by 69% and Europe and the Americas by 68%. The UNWTO (2021) states that the number of arrivals for 2020 is reflective of international tourism levels of thirty years ago. In tandem, the cruise industry experienced a 77-billion-dollar decline in global economic activity and a loss of 518,000 jobs (Cruise Lines International Association 2021). Among the most vulnerable in the current pandemic are Small Island Developing States (SIDS) with tourism accounting for '...30% of exports for the majority of SIDS and 80% for some' (UNWTO 2020b, 3).

Evidence from the fallout of the COVID-19 pandemic in 2020 on small economies reliant on the tourism industry is indicative of the importance of the sector to their economic well-being (Coke-Hamilton 2020). In the first four months of 2020, tourist arrivals to SIDS declined by 47% with an equivalent loss of approximately 7.5 million arrivals and in April 2020 alone, arrivals fell by an unprecedented 97% (UNWTO 2020a). In July 2020, Mulder (2020, 6) reporting on the impact of the COVID-19 pandemic on the tourism sector in Latin America and the Caribbean, predicted significant job loss in the tourism sector for Caribbean islands citing that, '...total employment may be reduced by 9 percentage points in the Caribbean... In the Caribbean, St. Lucia, the Bahamas, Antigua and Barbuda, St. Kitts and Nevis and Grenada may suffer a total employment loss of 15 percentage points or more...' At the time of this writing, these figures are still to be confirmed. The adverse effects of COVID-19 on the tourism sector on SIDS have impacted supporting sectors such as construction, agriculture and transport which would normally benefit from the circulation of the tourist dollar. This situation may appear paradoxical to many since SIDS in many cases have had the lowest incidence of COVID-19 cases and deaths but are experiencing the worst economic fallout compared to their larger counterparts. This is based largely on their dependency on a volatile tourism industry and their innate vulnerabilities.

Recovery Approaches

COVID-19, as a 'perfect storm' (Lew 2014; Ioannides and Gyimóthy 2020), has negatively affected economies of both the developed and developing world and has unexpectedly revealed inefficiencies within even more advanced systems. This current crisis propels tourism researchers to go back to the drawing board to review recovery frameworks expressly as it relates to SIDS given the high probability of unforeseeable future events (Miles and Shipway 2020; Ioannides and Gyimóthy 2020; Sigala 2020). Traditionally, tourism recovery is associated with the rebuilding of infrastructure, revitalization of tourist arrivals aided by relevant marketing and communications strategies (Pforr and Hosie 2008; Khazai et al. 2018). Tourism recovery due to COVID-19 is regarded differently from traditional tourism recovery '...largely due to the global scale and sheer complexity....' (Fakhruddin et al. 2020) and is being viewed as an opportunity to engage in radical transformation, innovation and regenerative measures (Brouder 2020; Cave and Dredge 2020) that are dynamic and adaptive (Bangwayo-Skeete and Skeete 2020; Hartman 2020). In specific reference to SIDS, Sheller (2020, 7) writes that it involves a '...re-think...of radical reconstruction and post-tourism sustainable development.'

However, Cave and Dredge (2020) state emphatically that 'new and alternative economic models in tourism"' needed in tourism recovery from 2020 and beyond may be stymied by the old way of doing things over several decades of growth. This position was also reflected by Sigala (2020, 117) who states '...the economic system and mindset contributing to the COVID-19 has also been guiding and shaping the COVID-19 response and recovery strategies of governments, institutions, businesses and people alike.' Hall et al. (2020) contend that it is unrealistic to expect any transformative approach to be employed in tourism development and that tourism will continue with business as usual with recovery hinging on 'economic and health interventions that include travel restrictions' with forms of tourism such as essential business travel, nature and VFR tourism leading in any recovery efforts. They argue that institutional and governmental capacity has been overwhelmed with a focus on saving lives, employing a basic restart of economies and saving jobs and businesses.

Some researchers view tourism recovery as a gradual process moving from one phase to the next. The World Bank Group (2020) details three phases of recovery in relation to COVID-19 which included new business models, capacity building and digitization. Phase 1 focuses on an immediate response (during the crisis and early recovery) needed to survive the crisis or to 'keep the lights on' and includes actions such as the provision of government tax relief, government grants, effective communication, subsidized training for tourism industry workers to forestall layoffs and acceptance of travel bans. Phase 2 stresses on short-term recovery including efforts to prepare for long-term recovery and involves the mobilization of digital platforms, developing a periodic survey and assessing the need of the tourism private sector, incentivizing business expansion and entrepreneurship, relaunching the destination through 'back in business' augmented promotion to key source markets, improving access to finance for registered tourism SMEs and strategizing for recovery. Finally, Phase 3 involves medium-term recovery including support for sector sustainability through greening, building awareness of tourism crisis management, strengthening coordination of public and private sector collaboration and direct public investment and conservation for jobs and sustainability. The aim being to ensure that the destination is in a better position prior to the crisis. These three phases weigh heavily on government intervention which is supported by Bangwayo-Skeete and Skeete (2020) where they indicate that Grenada was able to recover quicker than Barbados from the 2008 financial crisis due to adequate international, regional and government support, macroeconomic restructuring and political and social cohesion. The role of policymakers in the destination recovery was also elaborated on by Fakhruddin et al. (2020) who determined that responses could be effective or ineffective in the recovery process (Table 13.1).

Reddy et al. (2020) in their research proposed a conceptual framework for post-conflict tourism destination development noting that recovery is characterized by two distinct phases, the destination reorganization/normalization phase and the phoenix phase. They recommend the framework to destinations that have experienced a long-term shock and

Table 13.1 Destination recovery responses

Effective responses	*Ineffective responses*
Transparent governance, collaborative structures	Top-down governance, bureaucratic structures
Efficient and effective information dissemination	Lack of knowledge on how to disseminate information correctly
Modern information technologies and well-developed communication channels	Poor technology and fragmented communication channels
Dissemination of information to targeted population in a transparent manner, resulting in trust and engagement by the public	Inadequate/inconsistent information or misinformation, resulting in mistrust by the public
Strong community vigilance through public education and incentives	Weak community and lack of public education measures
Strong collaboration of major parties including city councils, citizens, and community volunteers	Lack of collaboration between major parties with lack of risk management integration into major (e.g., health, infrastructure, tourism, environment)
Evidence-based decision making, with the effective use of big data	Lack of data interoperability and meta data standardisation
Stringent hospital infection control measures, hygiene practices and use of personal protective equipment designating separate zones within the hospital or certain hospitals for infected patients only	Inadequate personal protective equipment and hygiene practices, no separation between the infected and non-infected patients
Continuing support during the lockdown	Lack of support to community in lockdown

Source Adapted fromFakhruddin et al. (2020)

who wish to re-engineer their tourism assets in the short term to recapture visitor numbers. Innovation, learning, transformation and coping capacity characterize redesign efforts and form the phoenix phase. They cited the development of narco tours as an example of the phoenix (recovery) phase in Colombia which they determined was created because of the destination's poor international reputation owing to drug related crime. Prior to this stage, the destination continues despite the crisis and also leveraging on relationships within the tourism destination system between local communities, tourism-related businesses and decision-makers. This position implies that consideration should be given to new tourism experiences and services as a main component of destination recovery (Lew et al. 2020; Brouder 2020; Cave and Dredge 2020).

Methodology

Six (6) Caribbean islands were selected for the study based on tourism revenue as a percentage of GDP and tourism as a share of total exports. The Bahamas, St. Lucia, Antigua and Barbuda, Barbados, British Virgin Islands and St. Kitts and Nevis were identified as some of those islands with tourism accounting for 50% or more of total exports. The study utilized qualitative document analysis (QDA) where various discourses were tracked over a specific timeframe by reviewing articles, press releases, reports, interview scripts, event programmes and other public records (Bowen 2009). As such, the author undertook '…skimming (superficial examination), reading (thorough examination), and interpretation…' (Bowen 2009, 32) of 80 press releases, 10 destination websites, 30 online news articles, 5 reports and 10 video interview/webinar scripts all sourced within March 2020 to March 2021. Following initial examination and interpretation, the information was collated in table form from which content analytic strategies were used to ascertain themes and patterns in relation to recovery approaches. Emerging from the content were three main themes, namely, national responses, tourism operation responses and marketing and communication responses (see Table 13.2).

Discussion

The literature highlights a clear divide between the traditional recovery process and what can be referred to as a transformational recovery process. Proponents of the latter (Fakhruddin et al. 2020; Brouder 2020) are encouraging island states to take this opportunity to revisit their starting context and engage in a new way of thinking with a focus on regenerative measures for the future. While it is broadly accepted that a transformational approach is necessary, the traditionalists (Hall et al. 2020) are of the view that it is the economic imperative that will guide decision-making at this time with minimal focus on innovation. These two viewpoints converge in the work of Reddy et al. (2020) whose conceptual framework demarcates a normalization phase (immediate economic interventions) and a phoenix phase (innovation and transformation). An analysis of the destination responses across the three (3) themes reveals that these island states are primarily in the normalization phase which aligns with phases I and II of the World Bank Group's (2020) recovery phases. This is not surprising, given the fact that the pandemic is just over a year in duration, so the focus is on saving lives and stabilizing the economy.

Table 13.2 Sample of Caribbean countries COVID-19 tourism recovery responses

Country	*National responses*	*Tourism operation responses*	*Marketing and communication responses*
Antigua and Barbuda	Borders closed to international travel Restrictions on public gatherings/Curfews imposed Establishment of multi-stakeholder tourism recovery taskforce Development of stimulus packages Phased re-opening of border—1st destination in the Caribbean to do this Quarantine measures/PCR Testing/Vaccination programme Commits to regional travel bubble	Health and safety protocols for the tourism sector established Establishment of Health and Safety certification programme 'Sun Sea Safe' for tourism businesses Training of tourism sector staff Provided list of approved accommodation facilities and transportation providers Utilization of new technologies e.g., a mobile app 'Explore Antigua' developed in partnership with a local company and features hotels, restaurants, bars, and popular tourist attractions around the island and is complete with a Google-integrated map for GPS navigation	Launch of social media marketing campaigns e.g., 'Message in the Sand', 'Your Space in the Sun' and 'Business on the Beach' campaigns Focus on domestic tourism e.g., 'Space Points Rewards Program' encourages residents of Antigua and Barbuda to dine-out Sales promotion initiatives e.g., wellness videos on social media highlights the destination as a destination to one's improve health; travel trade webinars in major source markets displaying the destination's key selling points; Your Space in the Sun' Sweepstakes offering a prize of free airfare and accommodation Development of new tourism products e.g., Launch of its Nomad Digital Residence (NDR) Programme, which allows persons to work from the twin-island up to two years with the new NDR visa Webinars e.g., Cruise tourism readiness webinar with the public entitled 'Inspiring the Restart of Cruise Tourism: Antigua & Barbuda Together' to encourage tourism and travel-related businesses to prepare for the relaunch of the local cruise tourism industry

(continued)

Table 13.2 (continued)

Country	*National responses*	*Tourism operation responses*	*Marketing and communication responses*
Bahamas	Borders closed to international travel Restrictions on public gatherings/Curfews imposed Establishment of multi-stakeholder tourism recovery taskforce Development of stimulus packages. Specifically, unemployment assistance is provided for self-employed tourism workers Phased re-opening of border Quarantine measures/PCR Testing/Rapid Antigen Test Vaccination programme/Health Visa required	Health and safety protocols for the tourism sector established Establishment of Health and Safety certification programme 'Clean and Pristine, Responding to Covid-19' for tourism businesses Training of tourism sector staff Utilization of new technologies—'The Islands of the Bahamas' App launched; Interactive island guide providing information to visitors on business operating hours, access to beaches etc. Integration of accommodation facilities into public health response e.g., Mandatory 14-day Vacation-In-Place (VIP) Experience at a hotel, private club or rented accommodations (like Airbnb), as well as on a private boat	Launch of social media marketing campaign e.g., 'Healthy Traveler Campaign' launched to encourage both visitors and residents to continue practicing all COVID-19 protocols Focus on domestic tourism e.g., Domestic tourism campaign to promote movement within the outer islands Provision of virtual experiences e.g., hosts a Virtual Romance Expo for consumers from all over North America titled 'From Bahamas with love' Sales promotion initiatives e.g., Out Island Air Credit promotion launched offering a US$250 credit to US & Canadian residents Development of new tourism products e.g., the Bahamas Extended Access Travel Stay (BEATS) program, a one-year residency permit for professionals and students wishing to work/study remotely from The Bahamas

(continued)

Table 13.2 (continued)

Country	*National responses*	*Tourism operation responses*	*Marketing and communication responses*
Barbados	Borders closed to international travel Restrictions on public gatherings/Curfews imposed Development of stimulus packages Establishment of tourism recovery taskforce Phased re-opening of border Quarantine measures/PCR Testing/Vaccination programme Commits to regional travel bubble	Health and safety protocols for the tourism sector established Training of tourism sector staff Provided list of approved accommodation facilities and transportation providers	Launch of social media/marketing campaign e.g., at the start of the pandemic Barbados reminds its visitors that the destination is awaiting their return under their 'Barbados Nice' campaign; social media campaign of testimonials of tourists visiting Barbados with the re-opening of the border or who have engaged in the remote work programme; these testimonials seek to reiterate how safe the destination is; 'Isleaway' campagain launched targeting regional visitors Sales promotion initiatives e.g., 'Barbados Cares' campaign is launched in the UK targeting 20 NHS heroes with a seven night trip to the island Development of new tourism products e.g., the 12 Month Barbados Welcome Stamp is launched as part of a new remote work programme allowing people to work remotely in Barbadas for a maximum of 12 months; 'Re-Imagine Tourism in Barbados' launched under the National Tourism Programme as a partnership between Barbados and the IDB to revitalize tourism offerings with specific focus on cultural heritage tourism Destination Updates e.g., introduced the Barbados Travel Guide Podcast-The Welcome Stamp Edition

(continued)

Table 13.2 (continued)

Country	*National responses*	*Tourism operation responses*	*Marketing and communication responses*
British Virgin Islands	Borders closed to international travel Restrictions on public gatherings/Curfews imposed Establishment of national recovery taskforce Development of stimulus packages Phased re-opening of border Quarantine measures/PCR Testing/Rapid Antigen Test Vaccination programme/	Health and safety protocols for the tourism sector established e.g., the 'BVI Tourism Reopening Guide' developed with relevant protocols Establishment of Health and Safety certification programme 'Gold Seal' for tourism businesses Training of tourism sector staff Provided list of approved accommodation facilities and transportation providers	Launch of social media /marketing campaign e.g., launches the 'Find yourself in the British Virgin Islands' and 'BVILOVE' campaigns Sales promotion initiatives e.g., Travelers can win 2 round trip airline tickets between St. Thomas, USVI and Tortola, BVI or San Juan, Puerto Rico and a 5-night stay at a villa as part of the 'find yourself' campaign; the destination partners with a private jet travel company to promote luxury travel; introduces the BVU Virtual Happy Hour to engage existing and potential visitors Provision of virtual experiences e.g., engages in a series of virtual roadshows titled 'Golden Virtual Roadshow' Destination Updates e.g., Launches a Travel Advisor Newsletter; develops a welcome to BVI COVID-19 brochure
St. Kitts and Nevis	Borders closed to international travel Restrictions on public gatherings/curfews imposed Development of stimulus packages Establishment of national recovery taskforce Phased re-opening of border Quarantine measures/PCR Testing/Vaccination programme Commits to regional travel bubble	Health and safety protocols for the tourism sector established under a Establishment of Health and Safety certification programme for tourism businesses 'Travel Approved Certificate'. Properties all receive the 'Travel Approved Seal' Training of tourism sector staff Provided list of approved accommodation facilities and transportation providers	Launch of social media /marketing campaign e.g., Nevis launches new adventure video to highlight the adventurous side of the island and declares it readiness for receives visitors; Nevis launches the 'Health Campaign' to promote safe travel to the destination; launch of 'Nothing Like a Nevian Holiday' Focus on domestic tourism e.g., Negotiated special hotel rates for locals Sales promotion initiatives e.g., "#RetieTheKnot' photo contest launched targeting couples in the USA to submit their favorite wedding picture to win a trip for 2 to the destination; Nevis launchs a video series competition, *#TasteofNevis*, a celebration of Nevisian culture

(continued)

Table 13.2 (continued)

Country	*National responses*	*Tourism operation responses*	*Marketing and communication responses*
		Integration of accommodation facilities into public health response e.g., Mandatory 14-day Vacation-In-Place (VIP) Experience at approved hotels	Marketing partnerships e.g., St. Kitts and Nevis, Saba, Statia, Sint Maarten, Anguilla and St. Barths formed the 'Caribbean Group of 8' joint marketing initiative to collectively raise the profile of each destination among travelers Development of new tourism products e.g., Extended stay/remote work programme introduced
St. Lucia	Borders closed to international travel Restrictions on public gatherings/Curfews imposed Development of stimulus packages Establishment of national recovery taskforce Phased re-opening of border Quarantine measures/PCR Testing/Vaccination programme Commits to regional travel bubble	Establishment of Health and Safety certification Training of tourism sector staff Focus on staff welfare e.g. launches SUPPORT SAINT LUCIA (donations campaign- funds will go directly towards on-the-ground support) Provided list of approved accommodation facilities and transportation providers Legislation e.g. a government tax named the "Tourism Levy' is introduced. Guests staying at registered accommodation facilities will pay a nightly levy. Funds will be used for tourism marketing and development	Launch of social media campaigns e.g., 'She is Saint Lucia…Let Her Inspire You'; 'She is Waiting' campaign, "Bubblecation" and '12 days of Christmas' campaigns Sales promotion initiatives e.g., partnered with American Airlines on a multi-faceted campaign featuring email marketing to more than 13 million AAdvantage members Webinars e.g., To further engage with trade, the UK SLTA team hosted regular webinars for agents and operators Destination updates e.g., 'It's in Our Nature' video speaks to solidarity and hope in the wake of the worldwide COVID-19 pandemic; rolls out its 'Saint Lucia Inspires' Blog Series; launches the first ever podcast series highlighting nature and wildlife, arts and culture, culinary delights, adventure and wellness of the destination

The national responses captured the governments' decisions on limiting the spread of the virus and sustaining the livelihoods of the locals. In terms of the former, the closure of the borders to international travel was an immediate response taken by all the destinations. However, from June 2020 onwards, there has been a phased reopening of the borders to international travel. The majority of the islands also committed to the establishment of a regional travel bubble with its neighbors. In all cases where entry to the islands was permitted, a negative PCR test was required and some level of quarantine was mandated. The next step for these islands is the roll out of the respective vaccination programmes. Coupled with the border closures, internally, there have been a series of restrictions on public gatherings and imposed curfews. With respect to the latter, all destinations established recovery committees or a task force and developed stimulus packages for their residents. It is noteworthy that the recovery committees comprised of a mix of public and private sector representatives which point to a high level of stakeholder engagement and support the need for stakeholder cohesion.

Alongside the national socio-economic interventions by the respective governments, the tourism sector in the respective islands embarked on a series of initiatives that focused on business operations, staff welfare and destination marketing strategy primarily premised on digital engagement. The common denominator for tourism businesses in these islands is the compliance with established health and safety protocols. Each island launched a Health and Safety Programme for tourism businesses. British Virgin Islands developed a 'Gold Seal' certification programme while St. Kitts and Nevis established a 'Travel Approved Certificate'. Antigua and Barbuda its certification programme 'Sun Sea Safe' for tourism businesses while the 'Clean and Pristine, Responding to Covid-19' programme was the certification title in The Bahamas. In addition to the local certification programmes, these islands have also been pursuing the Safe Travels Stamp granted by the World Travel and Tourism Council, an achievement already attained by The Bahamas, Barbados and St. Lucia. It is apparent from the findings that a key component of a post-COVID-19 tourism operating environment is well-defined health and safety protocols that are adhered to by tourism businesses. It was also noted that the well-being of residents is as important as that of the tourists as evidenced by the Support St Lucia Donation campaign.

A major challenge for SIDS during this pandemic is dealing with the travel restrictions and maintaining interest in the destination with the target audience. The destination messaging in the (6) territories has centered on hope and reminders about the destination, e.g., 'Your Space in the Sun' campaign in Antigua and Barbuda, 'Barbados Nice' in Barbados, 'Find Yourself in the British Virgin Islands' in the BVI, 'Nothing Like a Nevian Holiday' in Nevis and 'She is Waiting' in St. Lucia. This messaging is accompanied by a series of sales promotions including the Space Points Rewards Programme in Antigua and Barbuda, the Air Credit promotion in Bahamas, #RetieTheKnot photo contest in St. Kitts and Nevis. With the restrictions on international travel, attention is turned to the domestic market by promoting staycations with negotiated special local hotel rates in St. Kitts and Nevis and promotion of movement within the outer islands of the Bahamas. These marketing and communication initiatives are associated with the traditional tourism recovery approach that is squarely focused on preserving jobs in the tourism industry (Pforr and Hosie 2008; Khazai et al. 2018).

Glimpses of a favorable response to the call to reimagine (Ioannides and Gyimothy 2020), to innovate (Fakhruddin et al. 2020), to engage in the phoenix phase of innovation and transformation (Reddy et al. 2020) are seen in the development of 'new' tourism offerings through the leveraging of technology. The Bahamas and Antigua and Barbuda capitalized on the global shift to remote work by introducing the Nomad Digital Residence Programme and the Bahamas Extended Access Travel Stay programme, respectively. In like manner, Barbados launched a 12-month Welcome Stamp as part of a new remote work programme allowing persons to work for a maximum of 12 months. Both Antigua and Barbuda and the Bahamas launched mobile applications to allow for improved navigation of the destination. Barbados and the BVI joined The Bahamas and Antigua and Barbuda in hosting virtual tourism expos that targeted both the travel trade and the potential tourist. St. Lucia and Barbados employed additional communication technologies including podcast series, blogs, webinars and e-newsletters to engage the travel trade and potential tourists.

Tourism Resilience Implications

The COVID-19 pandemic has created a resilience conundrum for all countries, and not least for developing countries SIDS. For the first time, our often commented upon 'isolation status' has served us in relatively good stead, with virus and death numbers being much lower than that of countries in the Global North. According to the data, as of May 2, 2021, total confirmed cases among twenty Caribbean countries were 133,986 including 2,435 deaths compared to 152,534,452 cases globally including 3,198,528 deaths (WorldOMeter 2021). Still, these numbers have been potentially overwhelming for the health systems in many SIDS. The other side of the 'isolation argument' is that for many SIDS, our tourism overdependence has been laid bare by the coronavirus. Commenting on this, Grenade (2021, n.p) observes 'one of the *root* causes for the spread of the virus is global travel and tourism, which is also the *route* to recovery for Caribbean economies.' This is a profound reflection even as it reflects the state of our tourism overdependence as island states. Indeed, if then tourism is regarded by governments as the route to economic recovery, it is axiomatic that resilience strategies would privilege destination resilience and by extension business resilience, so that the crisis communication and promotional messages, the retraining of hospitality workers and the push by many to reopen borders, despite dissenting voices, are expected. Similar to Dahles' (2018) study of small tourism businesses in Indonesia, the data shows that pandemic business resilience strategies among Caribbean tourism businesses have largely centered on survival (absorption) and adaptation and less on innovation. Perhaps, this is likely to be the case for a while as many owners are understandably preoccupied with simply 'staying afloat' as suggested by Phase 1 of the World Bank's recovery steps.

The adaptive cycle proposed by Holling (2001) suggests that it is in the final phase of the cycle—reorganization—that the greatest potential exists for reimagination and innovation, a notion similar to the phoenix phase articulated by Reddy et al. (2020) and the transformative phase of Bene et al. (2012). The question becomes one of which entity will lead this resilience direction. It is unlikely that the private sector with their high capital investments would, and therefore it falls to governments of SIDS to engage in strategic transformative resilience visioning that would pursue other economic alternatives. Adopting a socio-ecological perspective that recognizes the interdependency in the global system is key. In this regard, regional and international partners, including the region's skilled diaspora, must be involved, in order to secure buy-in, technical assistance, knowledge transfer, funding and facilitation, all of which are seen as critical requirements in building a transformative agenda.

References

Bangwayo-Skeete, P.F., and Ryan W. Skeete. 2020. Modelling Tourism Resilience in Small Island States: A Tale of Two Countries. *Tourism Geographies*. https://doi.org/10.1080/14616688.2020.1750684.

Bene, C., R.G. Wood, A. Newsham, and M. Davies. 2012. Resilience: New Utopia or New Tyranny? Reflection About the Potentials and Limits of the Concept of Resilience in Relation to Vulnerability Reduction Programmes. *Institute of Development Studies Working Paper 2012* (405). https://doi.org/10.1111/j.2040-0209.2012.00405.x.

Birkmann, J., O. Cardona, M. Pelling, and D. Alexander. 2013. Framing Vulnerability, Risk and Societal Responses: The MOVE Framework. *Natural Hazards* 67: 193–211.

Bowen, G.A. 2009. Document Analysis as a Qualitative Research Method. *Qualitative Research Journal* 9 (2): 27–40.

Brouder, P. 2020. Reset Redux: Possible Evolutionary Pathways Towards the Transformation of tourism in a COVID-19 World. *Tourism Geographies* 22 (3): 484–490.

Cave, J., and D. Dredge. 2020. Regenerative Tourism Needs Diverse Economic Practices. *Tourism Geographies*. https://doi.org/10.1080/14616688.2020.1768434.

Coke-Hamilton, P. 2020. Impact of COVID-19 on Tourism in Small Island Developing States. *UNCTAD*, April 24, 2020. https://unctad.org/news/impact-covid-19-tourism-small-island-developing-states.

Cruise Line International Association. 2021. 2021 State of the Cruise Industry Outlook. Accessed February 10, 2021. https://cruising.org/-/media/research-updates/research/2021-state-of-the-cruise-industry_optimized.ashx

Dahles, Heidi. 2018. The Sustainability of Small Business Resilience: The Local Tourism Industry of Yogyakarta, Indonesia a Decade After the Crisis. In *Tourism, Resilience and Sustainability: Adapting to Social, Political and Economic Change*, ed. Joseph M. Cheer and Alan A. Lew, 149–164. Abingdon, Oxford: Routledge.

Everest-Phillips, M. 2014. *Small, so Simple?: Complexity in Small Island Developing States*. Singapore: UNDP Global Centre for Public Service Excellence.

Fakhruddin, B.S., K. Blanchard, and D. Ragupathy. 2020. Are We There Yet? The Transition from Response to Recovery for the COVID-19 Pandemic. *Progress in Disaster Science* 7: 100–102.

Grenade, Wendy. 2021. *The COVID-19 crisis: ReImagining New Pathways for the Caribbean*. Lecture delivered at 6th Carol Bristol Lecture Series. Hosted by the University of the West Indies, Open Campus, 30 March 2021.

Hall, C.M., D. Scott, and S. Gössling. 2020. Pandemics, Transformations and Tourism: Be Careful What You Wish for. *Tourism Geographies* 22 (3): 577–598.

Hartman, S. 2020. Adaptive Tourism Areas in Times of Change. *Annals of Tourism Research*. https://doi.org/10.1016/j.annals.2020.102987.

Holling, C.S. 2001. Understanding the Complexity of Economic, Ecological, and Social Systems. *Ecosystems* 4 (5): 390–405. https://doi.org/10.1007/s10021-001-0101-5.

Inter-American Development Bank. 2020. Caribbean Quarterly Bulletin: The Pandemic Saga Continues. Accessed January 20, 2021. https://publications.iadb.org/publications/english/document/Caribbean-Quarterly-Bulletin-Volume-9-Issue-2-July-2020.pdf.

Ioannides, D., and S. Gyimóthy. 2020. The COVID-19 Crisis as an Opportunity for Escaping the Unsustainable Global Tourism Path. *Tourism Geographies* 22 (3): 624–632.

Khazai, B., F. Mahdavian, and S. Platt. 2018. Tourism Recovery Scorecard (TOURS)–Benchmarking and Monitoring Progress on Disaster Recovery in Tourism Destinations. *International Journal of Disaster Risk Reduction* 27: 75–84.

Lew, Alan A. 2014. Scale, Change and Resilience in Community Tourism Planning. *Tourism Geographies* 16 (1): 14–22.

Lew, Alan A., Joseph M. Cheer, Michael Haywood, Patrick Brouder, and Noel B. Salazar. 2020. Visions of Travel and Tourism After the Global COVID-19 Transformation of 2020. *Tourism Geographies* 22 (3): 455–466.

Meddeb, R. 2020. How Can Small Islands Reimagine Tourism for a Green Recovery. *UNDP*, October 26, 2020. https://www.undp.org/content/undp/en/home/blog/2020/how-can-small-islands-reimagine-tourism-for-a-green-recovery.html.

Miles, L., and R. Shipway. 2020. Exploring the COVID-19 Pandemic as a Catalyst for Stimulating Future Research Agendas for Managing Crises and Disasters at International Sport Events. *Event Management* 24 (4): 537–552.

Mulder, N. 2020. *The Impact of the COVID-19 Pandemic on the Tourism Sector in Latin America and the Caribbean, and Options for a Sustainable and Resilient Recovery*. Santiago: UN. https://www.cepal.org/sites/default/files/publication/files/46502/S2000751_en.pdf.

OECD. 2018. *Making Development Co-operation Work for Small Island Developing States*. Paris: OECD Publishing.

Office of the High Representative for the Least Developed Countries, Landlocked Developing Countries and Small Island Developing States (UN-OHRLLS). 2011. Small Island Developing States Small Islands Big(Ger) Stakes. Accessed December 11, 2019. http://unohrlls.org/custom-content/uploads/2013/08/SIDS-Small-Islands-Bigger-Stakes.pdf.

Pforr, Christof, and Peter J. Hosie. 2008. Crisis Management in Tourism. *Journal of Travel & Tourism Marketing* 23 (2–4): 249–264.

Ramrattan, D. 2020. *Tourism-Fostering Development in Growth* [PowerPoint Slides]. Caribbean Development Bank.

Reddy, M.V., S.W. Boyd, and M. Nica. 2020. Towards a Post-Conflict Tourism Recovery Framework. *Annals of Tourism Research.* https://doi.org/10.1016/j.annals.2020.102940.

Sheller, M. 2020. Reconstructing Tourism in the Caribbean: Connecting Pandemic Recovery, Climate Resilience and Sustainable Tourism Through Mobility Justice. *Journal of Sustainable Tourism.* https://doi.org/10.1080/09669582.2020.1791141.

Sigala, M. 2020. Tourism and COVID-19: Impacts and Implications for Advancing and Resetting Industry and Research. *Journal of Business Research* 117: 312–321.

UNEP. 2014. *Emerging Issues for Small Island Developing States.* Accessed January 20, 2021. https://sustainabledevelopment.un.org/content/documents/2173emerging%20issues%20of%20sids.pdf.

United Nations World Tourism Organization. 2020a. *Impact Assessment of the Covid-19 Outbreak on International Tourism.* Accessed February 1, 2021. https://www.unwto.org/impact-assessment-of-the-covid-19-outbreak-on-international-tourism.

United Nations World Tourism Organization. 2020b. *UNWTO Briefing Note—Tourism and COVID-19, Issue 2. Tourism in SIDS—The Challenge of Sustaining Livelihoods in Times of COVID-19.* Madrid: UNWTO.

United Nations World Tourism Organization. 2021. *Worst Year in Tourism History with 1 Billion Fewer International Arrivals.* Accessed February 1, 2021. https://www.unwto.org/news/2020-worst-year-in-tourism-history-with-1-billion-fewer-international-arrivals#:~:text=Global%20tourism%20suffered%20its%20worst,World%20Tourism%20Organization%20(UNWTO).&text=The%20crisis%20has%20put%20between,small%20and%20medium%2Dsized%20enterprises.

World Bank. 2020. *Rebuilding Tourism Competitiveness: Tourism Response, Recovery and Resilience to the COVID-19 Crisis.* Washington, DC: World Bank.

WorldOMeter. 2021. Coronavirus Updates. Accessed May 5, 2021. https://www.worldometers.info.

CHAPTER 14

Conclusion

Sherma Roberts, Acolla Lewis-Cameron, and Leslie-Ann Jordan

Introduction

The preceding chapter explored the effects of the COVID-19 pandemic in Caribbean small island states and suggested that despite the profound effects of the pandemic on lives and livelihoods, Caribbean governments and private sector bodies continue to hold fast to tourism overdependence as the strategy for navigating their way out of this pandemic conundrum. This is concerning for the development of a sustainable Caribbean future, as it demonstrates that transformative resilience or engineering transformative capacities are not part of the policy agenda of our governments. Neither is it arguably part of the recommendations of international donor

S. Roberts (✉)
University of the West Indies, Cave Hill Campus, Barbados
e-mail: sherma.roberts@cavehill.uwi.edu

A. Lewis-Cameron · L.-A. Jordan
University of the West Indies, St. Augustine, Trinidad
e-mail: acolla.lewis-cameron@sta.uwi.edu

L.-A. Jordan
e-mail: leslie-ann.jordan-miller@sta.uwi.edu

A. Lewis-Cameron et al. (eds.), *Managing Crises in Tourism*,
https://doi.org/10.1007/978-3-030-80238-7_14

bodies from whom we receive 'resilience aid' and technical assistance. Axiomatically, the tourism private sector is also unwilling to consider a future in which tourism is not central to the respective national economic agendas. Interestingly, tourism resilience research also inadvertently adopts a posture where the pillar of economic resilience is fortified by destination and business resilience strategies. This is, despite the fact that we understand the many negative impacts created by overdependence, unmanaged tourism and the inequities that often exist in tourism in the developing world and in tourism dependent SIDS.

Where then does that leave us? The foreword and introductory chapter pose the question upon which all of the essays of this collection hang, that is, 'how is tourism resilience understood and practiced in Caribbean SIDS and what are the factors that inform, undermine or indeed redefine the sustainable resilience agenda for these territories?' Taken as a whole, the discussions reflect a strong adherence to businesses and destinations absorptive and adaptive resilience capacities in the face of a myriad number of vulnerability shocks. Emerging as well are the ways in which social, economic, community and environmental resilience strategies have been employed to ensure the continuation of the destination and livelihoods of persons directly and indirectly employed in the industry. These strategies of coping and survival should not be diminished in any way, as they have resulted more than often, in positive outcomes for island peoples. The question though is whether the tried and tested resilience actions are sufficient to deliver visible and significant quality of life improvements for residents of Caribbean SIDS? Further, in the context of a rapidly changing hazard landscape in which economic, social and environmental vulnerabilities collide, are the strategies of resilience sufficiently robust to ensure our survival much less our transformation?

Sheller (2020, 6) cautions 'a need to think politically about forms of vulnerability and resilience in relation to the tourism sector's recovery in the future.' She further elaborates:

> Rather than understanding recent disasters as posing a threat to tourism economies in the Caribbean, and seeking government investment or international loans to rebuild them, we should instead understand extractive forms of tourism as a threat to the creation of sustainable Caribbean economies, and we should take the current interruption of global travel as an opportunity to re-think forms of radical reconstruction and post-tourism sustainable development. (p. 7)

Unlike the recommendations put forward in the chapters, Sheller appears to be calling for a dismantling of the current tourism system. The more pertinent call in our view is for a *diversification* of Caribbean economies to break the overdependence strangle hold. In a similar vein, the foreword of this volume, penned by Baldacchino, asks us to consider *diversification within* and *diversification beyond* tourism, as possibilities for strengthening our economies. Alternatives such as those put forward by Srinivasan et al. (2021) are worthy of consideration and include shifting to more niche tourism products that are low-density and high value-added and that create local intersectoral linkages, diversifying to the 'blue economy,' shipping, fisheries and aquaculture, and mainstreaming technology in productive processes to reduce costs and improve efficiencies. Relatedly, The Caribbean Disaster Emergency Management Agency (CDEMA) framework (see Fig. 14.1) for resilience suggests that if we are to advance to transformative resilience a number of requirements must be in place, implemented and continuously evaluated. There is no romanticizing the notion of transformative resilience or 'bouncing forward quickly.' It is clear that it is likely to be a costly exercise and like sustainable tourism

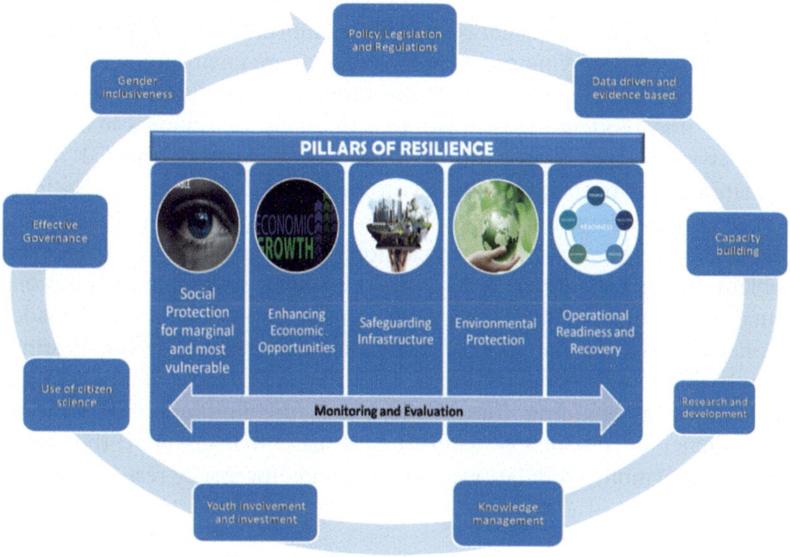

Fig. 14.1 Pillars of resilience (2018 *Source* CDEMA [, 4])

development, should be viewed on a continuum where incremental but steady progress is made.

Environmental Resilience

As highlighted by CDEMA, environmental degradation lies at the core of the challenges facing the Caribbean region. This is not surprising given the symbiotic relationship between Caribbean tourism and the natural environment. As seen in the case of St Lucia and Cozumel, the environment is under tremendous pressure owing to the intense tourism activity in these areas. Thus, environmental protection is a major pillar in building resilience in the Caribbean. In their framework CDEMA indicated that policy interventions in key areas such as promoting energy efficiency and renewable energy, strengthening the evidence base for decision making, incentivizing environmentally friendly practices are required to bring about transformational change.

The findings of the St. Lucia case study in Chapter 4 clearly show that legislation and regulations are also integral to supporting environmental resilience initiatives in the Caribbean. Without a legal framework, that includes monitoring, strict enforcement and stiff fines for breaches, the pillars of tourism resilience will be compromised. Regulations are needed to dictate carrying capacity in areas that are environmentally sensitive like coral reefs; and to treat with the use and disposal of plastics and other non-biodegradable waste that threaten the region's marine life and ecosystem. It is interesting to note that threats to the CT-MPAs originated from both cruise ship visitors to St. Lucia, as well as local tourism stakeholders like tour operators and fishermen. In this regard, building environmental resilience cannot be viewed simply as a 'tourism problem' but as a challenge that has wider societal, economic and environmental implications, and it is this approach that should frame the broader national agenda. In the case of Cozumel, it is clear that the destination is making some progress, albeit slow, in its environmental protection agenda. Notwithstanding these deficiencies, the findings also point to good practice in the areas of tourism seasonality management; the establishment of planning regulations and crisis and emergency response plans; the identification of environmental risks; and the protection of sensitive environments.

A major vulnerability shock for Caribbean islands is the frequently occurring hurricanes. Every year hurricanes wreak havoc on the fragile systems of the Caribbean. Although they are an inevitable hazard for the region, hurricanes still pose a formidable challenge for destinations in implementing relevant resilience strategies. The resilience actions of both Dominica and Puerto Rico post Hurricane Maria were examined in Chapters 6 and 7 respectively. Dominica has taken the bold step to move towards becoming the first 'hurricane-proof country' or the 'world's first climate-resilient island nation. At the core of this move was the development of a National Resilience Development Strategy to 2030 that incorporates disaster risk management mechanisms in tourism related facilities, disaster risk management training for stakeholders and capacity building in climate resiliency and disaster preparedness for industry stakeholders. The clear thrust for Dominica is the creation of an enabling environment at the macro level, alongside capacity development among industry stakeholders and local communities. The implementation of this strategy over the coming years will certainly set Dominica on a clear pathway for disaster resilience, acknowledging that success depends in large measure on the cooperation and partnership of diverse stakeholders.

Socio-Cultural Resilience

While the case study on Dominica took a macro perspective, the authors in Chapter 7 took a bottom-up approach and examined the resilience actions of a distinct group of industry stakeholders who were directly impacted by Hurricane Maria. The Puerto Rico case study puts the spotlight on industry stakeholders, adventure tour operators who in this case are the key players in disaster resilience. The authors highlighted important attributes of the adventure operators that were critical to the overall community resilience. They were adaptable, willing to take risks and had strong financial and social capital to support them. They were focused on strengthening the industry and their vision for adventure recreation in Puerto Rico was undiminished despite Maria's impact. In fact, Hurricane Maria was also a catalyst for change in terms of business expansion and product development. In the words of the authors, 'resilient community members drive community resilience.' What is clear from the post Hurricane Maria responses of both Dominica and Puerto Rico is the need for partnership. The government's creation of an enabling environment must be met with resilient stakeholders, who see the value in a collaborative approach.

The case study of Tobago's North–East UNESCO designation in Chapter 8 also detailed the increased public–private sector collaboration between the State and CSOs actors between 2014 and 2020. This trifecta of stakeholder management, partnership and collaboration created the level of trust that was needed to achieve consensus on the most appropriate path towards sustainable development, resilient and responsible tourism and conservation of ecosystem services in Tobago. The authors also mentioned the establishment of a Steering Committee which was the longest collaboration between state and non-state actors regarding conservation and sustainable natural resource management in the history of Tobago. It also signalled a transition from a colonial top-down decision and policy-making approach to a stakeholder participatory approach, which has been identified as one of the best practices for facilitating tourism resilience.

The Grenada case study (Chapter 9) underscores the importance and role of technological innovation and digital transformation in advancing the region's efforts to re-imagine, diversify and reset the tourism and travel industry post-pandemic. Integrating technology into the region's product development strategy will encourage the introduction of new tourism experiences such as leisure fishing, maritime archaeology and virtual exploration. This allows the region to build capacity while diversifying into other sustainable industries. Using the existing technology also encourages greater community involvement and participation in product development using social media like Twitter, TikTok, Facebook and Instagram to interact, educate and engage with potential future travelers, as well as local residents.

Economic Resilience

One of the (5) pillars proposed in the Figure 14.1, for progressing to a more resilient development in the Caribbean region, is enhancing economic opportunity. According to CDEMA, this must be based on the creation of an appropriate enabling environment, flexibility in social protection systems, incentivizing the business sector including Small and Medium Sized Enterprises and creating spaces for entrepreneurship and innovation. It is the latter notion of incentivization, entrepreneurship and innovation that occupy the attention of the authors of Chapters 10 and 11. Balkaran and David argue that a most vulnerable group in the face of disasters is the MSMEs as by virtue of their sheer small scale they are unable to bounce back to business as usual in quick time. The reality is, these businesses make up more than 50% of the businesses in the Caribbean and thus their survival is intrinsically related to the overall sustainability and survivability of the islands. Business continuity is paramount and therefore business incentivization schemes must form an integral part of economic resilience for island states. However, the authors warn of a tendency among MSMEs to be over-reliant on the public purse in times of a crisis, so governments need to be aware of this proclivity and manage the process efficiently.

One of the underlying objectives of enhancing economic opportunity is to increase the multiplier effect throughout the economy by facilitating effective forward and backward linkages among industries. As discussed in Chapter 2, a debilitating factor for Caribbean economies is the over-reliance on imports for the operations of the industry, primarily the accommodation sector. The deleterious effects of this overdependence are magnified in times of a global crisis when global supply chains are interrupted. While in many instances some Caribbean islands have no choice but to rely heavily on foreign imports, there are opportunities to decrease this reliance as seen in the case of Jamaica and the agriculture—tourism nexus. From their research, the authors surmise that there is a willingness among the accommodation establishments to use local farmers as their primary source of agriculture produce. However, the strengthening of the relationship between the establishments and the farmers requires the addressing of the consistency and quality of the produce and determining how leveraging existing institutional infrastructure can enable the farmer to understand the needs of the tourism industry.

A common thread throughout the majority of these chapters is the value of collaboration and public–private sector partnerships in order to create an enabling environment for tourism resilience. Smith and Jordan's (Chapter 12) research findings documented the benefits that rural communities can derive from pooling their financial resources, knowledge, marketing efforts and sustainable initiatives. To a large extent, the success of resilience strategies depends on partnerships between NGOs, public sector entities and civil society. While Castara's model is not void of challenges, it certainly presents an alternative to smaller communities that are trying to respond to economic shocks by finding sustainable ways to pivot.

Final Thoughts

In closing, three lessons are worth noting. One is that resilience is an integrated and holistic concept that favors interdependency in the system. In other words, for SIDS it is disadvantageous to focus attention solely on disaster resilience *or* economic resilience at any one time, as all of these dimensions intimately intersect with each other. There is also a macro interdependency perspective that is often overlooked in our quest for building resilient islands. Nowhere has this been more evident than in the COVID-19 pandemic, where the globalized nature of travel and tourism has led to the virus reaching far off often 'isolated' islands. A multisectoral, multi-level approach to resilience is therefore crucial in our search for solutions to reduce exposure and mitigate vulnerability.

Two, we propose that if the Caribbean is to progress towards transformative resilience, crime—local and transnational, must be frontally and cooperatively dealt with. In a 2021 crime index, four Caribbean island states were numbered in the top 20 among a total of 135 countries (www.numbeo.com). Reporting on corruption levels among 180 countries, Transparency International (2021) data revealed that 3 Caribbean countries were in the first quartile, 3 in the second and 1 in the third quartile, where the first quartile is least corrupt and the fourth is the most corrupt. While the Caribbean Community must be lauded for the early establishment of the Regional Security System that is at the forefront of a collective response to security threats, there needs to be a clearer articulation by SIDS that stability is a critical requirement for resilience (Bene et al. 2014). In this regard, resilience must go beyond the economic, environmental and social spheres to speak to wider systemic issues that are likely to undermine the very transformative process.

Three, is that if capacity building for all the dimensions of resilience is not mainstreamed throughout a broader national sustainable development vision, then the social, environmental, economic and institutional gains made in times of abundance can be lost due to one shock event. According to venerable Caribbean scholars, the capacity required among Caribbean small states to secure a more sustainable and resilient future lies in:

> fundamentals such as the development of human resources, institutional capacities and the creation of technological assets which constitute to the building blocks of development. Effective governance is part of the overall

> human development equation. It will be important therefore for the region to review existing constitutional arrangements and political processes. The role of the creative imagination in Caribbean societies [must be seen] as one of the most valuable assets of the region in its efforts to cope with the challenges of the twenty-first century. (Hall and Benn 2000, 588–589)

The focus then should not be on strengthening resilience for its own sake but in securing a sustainable future for all of society. In the case of islands with its tropes of vulnerabilities, this synergistic imperative should be even more compelling.

References

Béné, Christophe, Andrew Newsham, Mark Davies, Martina Ulrichs, and Rachel Godfrey-Wood. 2014. Review Article: Resilience, Poverty and Development. *Journal of International Development* 26 (5): 598–623. https://doi.org/10.1002/jid.2992.

CDEMA, and CARICOM. 2018. "Building a Caribbean Pathway for Disaster Resilience in the Caribbean Community (CARICOM)." Barbados: Bridgetown, July 2018. https://www.cdema.org/cdm11/Building_A_Caribbean_Pathway_For_Disaster_Resilience_In_The_Caribbean_Community_(CARICOM)_FINAL.pdf.

Hall, Kenneth, and Denis Benn. 2000. Conclusion. In *Contending with Destiny: The Caribbean in the 21st Century*, ed. Kenneth Hall and Denis Benn, 587–589. Kingston, Jamaica: Ian Randle Publishers.

Numbeo. Caribbean: Crime Index by Country 2018. Numbeo—Cost of Living, 2021. https://www.numbeo.com/crime/rankings_by_country.jsp?title=2018®ion=029.

Sheller, Mimi. 2020. "Reconstructing Tourism in the Caribbean: Connecting Pandemic Recovery, Climate Resilience and Sustainable Tourism Through Mobility Justice." *Journal of Sustainable Tourism* 29 (2): 1–14. https://doi.org/10.1080/09669582.2020.1791141.

Srinivasan, Krishna, Sònia Muñoz, and Ding Ding. 2021. How the Caribbean Can Avoid Becoming a COVID-19 Long-Hauler. IMF, March 12. https://www.imf.org/en/News/Articles/2021/03/11/na031221-how-the-caribbean-can-avoid-becoming-a-covid-19-long-hauler.

Index

A
adventure tourism, 121, 135, 136, 138, 140, 141, 144, 147–151
Agri-Linkages Exchange (ALEX), 223, 236
Antigua & Barbuda, vi, 272, 278
Aruba, xi, 22, 43

B
backward linkage, 225
Barbados, vi, 18, 19, 23, 38, 43, 44, 49, 51, 55, 76, 79, 272, 275, 277, 280, 283, 284
Barbuda, 18, 45, 198, 273, 277, 278, 283, 284
biodiversity, 67, 68, 81, 84, 156, 159, 162, 165, 168, 169
bouncing back, vii, viii, 56
British Virgin Islands (BVI), 18, 45, 198, 272, 277, 281, 283, 284
business ecosystems, 177, 180, 182, 184
business resilience, 53, 150, 285, 290

C
capacity building, xxii, xxvii, 49, 122, 263, 275, 293, 296
Caribbean Community (CARICOM), 46
Caribbean Disaster Emergency Management Agency (CDEMA), 34, 291
Caribbean environment, 25
Caribbean tourism, xxi, xxii, 16, 24, 25, 71, 177, 178, 180, 285
Caribbean Tourism Organization (CTO), xxii, 135, 176
carrying capacity, xxii, 68, 70, 71, 76–79, 81–83, 98, 186, 254, 264, 292
Castara, 159, 160, 242–244, 246–258, 260–262, 264
Castara Tourism Development Association (CTDA), 242, 247
Cayman Islands, 25, 44
Charlotteville, 159, 160, 165, 166
Citizenship by Investment (CBI), 120

A. Lewis-Cameron et al. (eds.), *Managing Crises in Tourism*,
https://doi.org/10.1007/978-3-030-80238-7

climate change, 8, 25, 35, 36, 42, 48, 77, 99, 100, 111, 151, 156, 158, 198, 200, 214
collaborative approach, 125, 243, 245, 248, 253
collaborative economy, 178
collaborative entrepreneurship, 243, 245–247, 255
community resilience, 34, 35, 48, 53, 56, 136–138, 144, 150, 246, 247, 293
Corporate Social Responsibility (CSR), 179
COVID-19, viii, x, xii, 16, 110, 115, 126, 181, 182, 279, 281–283
Cozumel, 8, 89, 90, 92–106, 292
creative industries, xi, 177
crisis communications, xxii, 115, 123
crisis management, ix, 103, 207, 275
cruise industry, 71–73, 185, 273
cruise ship, 25, 26, 71, 89, 111, 159
cruise ship tourism, 160
cruise tourism, xxii, 67, 68, 71, 72, 76, 77, 84, 92, 185, 278
Cuba, 14, 23, 25, 44, 100, 228
cultural tourism, 37, 100

D
dependency, 15, 16, 21, 22, 27, 55, 274
destination's resilience, 125
digital technology, 179, 180, 189
disaster management, 114, 120, 124
disaster planning, 114, 115, 120, 125
disaster risk reduction (DRR), 112, 199
disasters, vi, xxi, 13, 15, 16, 25, 27, 42, 45, 46, 109, 110, 112, 114, 124, 125, 135–137, 139, 140, 146, 148–150, 181, 199, 204, 208, 247
diversification, ix, xi, 15, 116, 147, 176, 180, 183, 189, 243, 247
Diversification strategies, xi, 52
Dominica, xxii, 5, 8, 24, 25, 44, 46, 110, 115, 117–120, 122–126, 138, 198, 293
Dominican Republic, 14

E
earthquakes, ix, 47, 109, 111, 114, 139, 197, 247
ecological carrying capacity, 68, 99
ecological resilience, 34, 54
economic crises, 8, 156, 176, 181
economic dependence on tourism, 18
economic shocks, vi, 43, 46, 176, 181, 247
educational tourism, 162, 169
enabling environment, 170, 293, 294
entrepreneurship, xxii, 46, 145, 243–247, 252, 262–264, 275, 294
environmental impact assessment (EIA), 76
environmental management systems, 102
environmental risks, 103, 292
environmental sustainability, 90, 158
environmental vulnerability, 4
evolutionary resilience, 34
exogenous shocks, 25, 34, 46, 48
external shocks, vii, 4, 8, 9, 13, 27, 45, 109, 156, 157, 162, 272

F
foreign exchange, 14, 18, 19, 22, 23, 34, 38, 168, 198, 200, 221, 226
forward linkage, 226

G
governance, xxii, 27, 39, 40, 43, 45, 46, 55, 106, 115, 116, 125, 126, 137, 158, 165, 250, 253, 296
Grenada, vi, xi, 8, 18, 44, 46, 55, 175–177, 180, 181, 183–189, 209, 273, 275
Grenada Tourism Authority (GTA), 187

H
Haiti, 14, 44
hazards, 34, 47, 48, 110, 112, 125, 136, 138, 139, 150, 197, 198, 200, 208, 214
Hurricane Andrew, 39
Hurricane Gilbert, 37, 39
Hurricane Irma, 41, 45, 141, 198
Hurricane Maria, 8, 25, 44, 46, 110, 118–121, 124, 135, 141–151, 198, 208, 293
hurricanes, ix, 15, 39, 103, 109, 111, 114, 139, 147, 150, 151, 156, 199, 200, 247

I
innovation, xxvii, 38, 55, 116, 124, 138, 149, 180, 185, 187, 190, 264, 274, 277, 284, 285, 294
institutional arrangements, 84, 214
institutional arrangements for tourism, 171
institutional capacity, 16, 256
institutional structures. *See* institutional arrangements
internal shocks, 37, 116, 176
International Monetary Fund (IMF), 20, 52

J
Jamaica, vi, xi, xxii, xxiii, 9, 14, 18, 19, 23–26, 39, 44, 200, 201, 206, 209, 212, 213, 222–224, 226, 227, 229, 234–236, 295

L
livelihood(s), xxi, 35, 56, 90, 92, 105, 106, 164, 183, 197–200, 202–204, 206, 209, 213, 243, 244, 272, 283, 289

M
Man and the Biosphere (MaB) Programme, 104, 155
Marine protected areas (MPAs), 68
Micro, Small and Medium Enterprises (MSME's), 202
Montserrat, vi, 44
multiplier effect. *See* multiplier theory
multiplier theory, 224

N
National Trust of Trinidad and Tobago, 165
natural hazards, 111, 138, 200
North-East Tobago, 8, 156, 157, 159–170

O
Office of Disaster Preparedness and Emergency Management (OPDEM), 213
overtourism, xi, 92

P
Participatory approaches, 213

partnerships, 90, 105, 106, 185, 212, 213, 246, 250, 282. *See also* public-private sector collaboration
political will, 78, 170
private and public sector collaboration, 126
public-private sector collaboration, 170, 294
Puerto Rico, xxiii, 8, 44, 135, 136, 140–151, 208, 281, 293

R

recovery plans, 115, 138
recovery strategies, 6, 8, 9, 121, 274
re-imaging, 113, 124, 178
resilience, vii–ix, xxi, xxii, 27, 34–42, 45–49, 51–55, 68, 70–72, 78–84, 90–92, 94, 110, 113, 116, 117, 120, 121, 124–126, 136–139, 141, 144–151, 156–158, 162, 169, 176–178, 180, 181, 183, 190, 198, 200, 204, 206, 208, 211, 213, 214, 222, 224, 229, 236, 245, 247, 250, 256, 257, 261, 265, 272, 284, 285
resilience-based framework, 116
resilience efforts, 51, 126
resilience strategies, 37, 45, 49, 51, 285
responsible tourism, xxvii, 90, 105, 162, 166, 170, 246, 247, 294
role of tourism, 13, 18, 162, 244
rural communities, 244, 245
rural tourism development, 244

S

service economy, 175
SIDS. *See* Small Island Developing States
Small Island Developing States (SIDS), 4, 13, 109, 156, 157, 181, 198, 273
small island states (SIS), vi, vii, ix, 14, 16, 34, 42, 43, 289
small island territories, x
socio-cultural sustainability, 158, 253
socio-ecological resilience, 34
Soufriere Marine Management Authority (SMMA), 78
Sphere of Tourism Resilience framework, 117
Stakeholder Collaboration, 250
St. Kitts and Nevis, 41, 272, 273, 277, 282–284
St. Lucia, vi, 8, 18, 70, 72–77, 79, 80, 83, 84, 209, 272, 273, 277, 282–284, 292
St. Martin, 41, 42, 45
St Vincent and the Grenadines (SVG), v
sustainable development, xxii, xxvii, 13, 26, 42, 80, 156, 158, 161, 163, 166–170, 176, 183, 188, 205, 262, 274, 290, 294, 296
sustainable tourism, xii, xxii, xxv, 16, 53, 54, 70, 91, 92, 97, 102, 138, 141, 158, 163, 170, 171, 189, 212, 245, 247, 253
SVG. *See* St. Vincent and the Grenadines

T

technological advancements, 177
technological innovation, 52, 179, 187, 244
technological transformation. *See* technological innovation
The Adaptive Cycle Model, 54
The Bahamas, 23, 26, 39, 52, 112, 272, 273, 277, 279, 284

Tobago, ix, xxv, xxvii, 26, 155–171, 201, 204, 206, 209, 213, 241–243, 247, 251, 253, 259, 263, 265, 294
Tobago House of Assembly (THA), 156, 247, 251
tourism-agriculture relationship, 224, 227, 228
Tourism Area Life Cycle, 54
Tourism Crisis and Disaster Management Framework, 115, 122
tourism-dependent islands, 126
tourism entrepreneurs, 244, 245
tourism product, 24, 147, 159, 162, 177, 182, 183, 185, 245, 251, 262
tourism resilience, 5–7, 9, 34, 52, 54, 55, 116, 151, 157, 177, 284, 290, 292
Training and Development, 250, 260
transformative agenda, 285
transformative resilience, 9, 39, 40, 42, 51, 285, 289, 291, 296
Trinidad and Tobago, ix, 25, 49, 156, 163, 165–169, 206, 209, 211, 241, 243, 253
tropical cyclones, 198, 200, 204

U
United Nations World Tourism Organization (UNWTO), 18

V
value chain approach, 226, 235
visitor management system, 98
vulnerability, vii, viii, xii, xxi, 13–15, 34, 35, 38, 41–43, 46–49, 52, 69, 111, 112, 137, 138, 140, 148, 179, 189, 198–200, 204, 206
Vulnerability-Resilience Nexus, 47

W
waste management, 76, 91, 94, 95, 105, 170
World Travel and Tourism Council (WTTC), 17, 141, 241, 283

The manufacturer's authorised representative in the EU is Springer Nature Customer Service Centre GmbH, Europaplatz 3, 69115 Heidelberg, Germany. If you have any concerns regarding our products, please contact ProductSafety@springernature.com

Printed and bound by CPI Group (UK) Ltd, Croydon, CR0 4YY
30/09/2026
02225452-0002